Lecture Notes in Mathematics

T0253818

Editors-in-Chief:
J.-M. Morel, Cachan
B. Teissier, Paris

Advisory Board:
Camillo De Lellis (Zürich)
Mario di Bernardo (Bristol)
Alessio Figalli (Austin)
Davar Khoshnevisan (Salt Lake City)
Ioannis Kontoyiannis (Athens)
Gabor Lugosi (Barcelona)
Mark Podolskij (Heidelberg)
Sylvia Serfaty (Paris and NY)
Catharina Stroppel (Bonn)
Anna Wienhard (Heidelberg)

More information about this series at
http://www.springer.com/series/304

Hatice Boylan

Jacobi Forms, Finite Quadratic Modules and Weil Representations over Number Fields

Springer

Hatice Boylan
Matematik Bölümü
İstanbul Üniversitesi
İstanbul, Turkey

ISBN 978-3-319-12915-0 ISBN 978-3-319-12916-7 (eBook)
DOI 10.1007/978-3-319-12916-7
Springer Cham Heidelberg New York Dordrecht London

Lecture Notes in Mathematics ISSN print edition: 0075-8434
ISSN electronic edition: 1617-9692

Library of Congress Control Number: 2014957642

Mathematics Subject Classification (2010): 11F50, 11F27

Printed on acid-free paper

Springer is part of Springer Science+Business Media (www.springer.com)

To my beloved father, Mustafa Boylan

Foreword

The current research monograph presents a breakthrough in at least three ways. Firstly, it introduces the simple but striking idea that the index of a Jacobi form over a totally real number field should be a lattice of rank one over the ring of integers rather than a number. The classical theory of Jacobi forms over the rational numbers uses positive integers for the index. Accordingly the first attempts to extend this theory to number fields tried to define the index by totally positive numbers in the different of the field. It soon turned out that this is too restrictive to obtain a satisfactory theory. However, if one views the index of classical Jacobi forms as (one half of) the Gram matrix of a rank one lattice over the rational integers, it becomes clear why, for general number fields, scalar indices will not suffice to capture all Jacobi forms. Indeed, as soon as the ring of integers of a given number field is no longer a principal ideal domain, we lose the one-to-one correspondence between lattices of rank one and numbers. The missing right notion of index blocked for a long time the research on Jacobi forms over number fields. As shown in this monograph, the consequent use of lattices as indices leads finally to a smooth and consistent theory.

Secondly, the development of a theory of Jacobi forms over number fields was also blocked by the lack of concrete or interesting examples. Again this monograph breaks this spell. It shows that there are indeed interesting examples. More precisely, it gives us in the last chapter a complete description of all Jacobi forms of singular weight on the full Hilbert modular group over any given totally real number field. These new functions generalize to number fields the classical Jacobi theta function which occurs in the Jacobi triple product, and which is essentially equal to the Weierstrass sigma function. One might expect that Boylan's theta functions will play a similar important role in the theory of Hilbert modular forms, the theory of abelian varieties or algebraic number theory as the classical Jacobi theta function, or equivalently, the Weierstrass sigma function.

Finally, indispensable for the study of Jacobi forms over the rationals is their realization as vector-valued modular forms. These modular forms take values in Weil representations. Again the theory of Jacobi forms over number fields was blocked since the corresponding objects for number fields were only vaguely known as part of Weil's general and abstract theory of what is known nowadays as Weil representations. For concrete considerations the Weil representations of Hilbert modular groups were apparently considered as too complicated. Again this research monograph surprises by showing that this is not at all true. It develops from scratch an appealing complete theory of finite quadratic modules and their associated Weil representations over arbitrary (not necessarily totally real) number fields. It describes the decomposition into irreducible parts of those Weil representations which are important for the Jacobi forms considered in this monograph. This theory alone provides already a valuable and indispensable tool for future research not only on Jacobi forms but also for the representation theory of Hilbert modular groups.

The theory of Jacobi forms over number fields is far from being at a stage as the corresponding theory over the rationals. However, the beauty of this theory already glimpses through if one looks at its basics and the concrete examples as they are presented in this book. This monograph will serve well as the cornerstone for building up a complete arithmetic theory of the newly introduced Jacobi forms. Indeed, there is currently already various work in progress.

In [SS14] the authors calculate the dimension of the spaces of vector-valued Hilbert modular forms with special emphasis on deriving explicit formulas for the dimensions of spaces of Jacobi forms over number fields of weight greater than 2. The approach is based on a general Eichler–Selberg trace formula for vector-valued Hilbert modular forms, and on the theory of Weil representations and their connection to Jacobi forms as developed in this monograph. The article [BHS14] determines the structure of the ring of Jacobi forms over $\mathbb{Q}(\sqrt{5})$ as module over the ring of Hilbert modular forms. The research project [SW14] aims to develop the theory of Hecke operators for Jacobi forms over number fields, and to study the connection of this new type of Jacobi forms to Siegel–Hilbert modular forms. The article [Boy14] calculates the Fourier coefficients of Jacobi Eisenstein series over number fields and gives thereby the first concrete examples for a deep arithmetic connection between the new Jacobi forms and Hilbert modular forms. In [BS14b] the authors plan to summarize the results of these (and possibly other) research activities, to give further explicit examples of liftings from Jacobi forms over number fields to Hilbert modular forms, and, in particular, to present a complete corresponding Hecke theory.

The main interest for constructing a theory for Jacobi forms over number fields arose from the fact that we expect several deep results from the theory of elliptic modular forms and Jacobi forms over $\mathbb{Q}$ to hold true for the number field case too. In particular, we expect liftings from Jacobi forms over number fields to Hilbert modular forms. Moreover, the Fourier coefficients of the Jacobi forms should encode the vanishing at the critical point of twisted L-functions associated with Hilbert modular forms. This is, in particular, interesting in the context of a generalized Birch and Swinnerton–Dyer conjecture for elliptic curves over number fields.

This monograph is an important step towards such a theory. It opens the door to a new and fascinating world of so far unseen functions. I hope that it will stimulate more researchers to follow this invitation to a new exciting subject.

Bonn, Germany
September 2014

Nils-Peter Skoruppa

Preface

In analogy to the theory of classical Jacobi forms which has proven to have various important applications ranging from number theory to physics, we develop in this research monograph a theory of Jacobi forms over arbitrary totally real number fields. However, we concentrate here mainly on the connection of such Jacobi forms and the theory of Weil representations, leaving out important topics like Hecke theory and liftings to Hilbert modular forms, which still have to be developed. We hope to come back to those topics in later publications, but that the present work stimulates already further interest in this rich new theory. Here, we develop, first of all, a theory of finite quadratic modules over number fields and their associated Weil representations. Next we develop in detail the basics of the theory of Jacobi forms over number fields and the connection to Weil representations. As a main application of our theory, we are able to describe explicitly all singular Jacobi forms over arbitrary totally real number fields whose indices have rank one. We expect that these singular Jacobi forms play a similar important role in this newly founded theory of Jacobi forms over number fields as the Weierstrass sigma function does in the classical theory of Jacobi forms.

I thank the Max-Planck Institute for Mathematics in Bonn for its hospitality and the beautiful research environment which they provided during the year 2013 when I was working on finalizing this book and on related topics. I also thank İstanbul Üniversitesi for allowing me to spend the year 2013 in Bonn. I thank the Mathematical Sciences Research Institute in Berkeley for hosting me in spring 2011 at a still early stage of this research project, and Tongji University in Shanghai and Chennai Mathematical Institute for giving me the possibility to lecture in fall 2009 on various very early results on Jacobi forms over number fields. Finally, I thank Nils-Peter Skoruppa for having me introduced to this fascinating subject, for showing constantly interest in my work, and for being always ready to share his profound vision of mathematics. Last but not least, I thank all those who cannot all be named here explicitly without omissions but have been influential for my career.

Bonn, Germany and İstanbul, Turkey
November 2013

Hatice Boylan

Preface

Contents

Introduction

A Jacobi form of weight k and index m (both half integral) on the full modular group $\mathrm{SL}(2,\mathbb{Z})$ is a holomorphic function $\phi(\tau,z)$ on the product $\mathbb{H}\times\mathbb{C}$ of the complex upper half plane $\mathbb{H}$ with the set of complex numbers $\mathbb{C}$ such that $\psi(\tau,x,y) := \phi(\tau,x\tau+y)\,e^{2\pi i m x^2\tau}$ satisfies the following properties:

(i) The function $\psi(\tau,x,y)$ is quasi-periodic in the real variables x and y with period 1.
(ii) For fixed rational x, y, the map $\tau\mapsto\psi(\tau,x,y)$ defines an elliptic modular form of weight k (possibly with character) on the principal congruence subgroup $\Gamma(a)$ of $\mathrm{SL}(2,\mathbb{Z})$, where a denotes the square of the least common multiple of the denominators of x and y.

The first property implies that, for fixed τ, the map $z\mapsto\phi(\tau,z)$ defines a theta function (a holomorphic section of a line bundle) on the elliptic curve $\mathbb{C}/(\mathbb{Z}\tau+\mathbb{Z})$. If we study n-dimensional abelian varieties whose endomorphism ring contains the ring of integers $\mathscr{O}$ of a totally real number field K of degree n over $\mathbb{Q}$, then we find naturally analogs of Jacobi forms. We will call them Jacobi forms over the number field K. A careful analysis shows, however, that we have to replace the index m by a totally positive definite integral $\mathscr{O}$-lattice of rank one. Such a lattice can always be represented by a pair $(\mathfrak{c},\omega)$, where $\mathfrak{c}$ is a fractional $\mathscr{O}$-ideal and ω a totally positive element in K such that $\mathfrak{c}^2\omega$ is contained in the inverse different of K (see Proposition 3.10). If K is the field of rational numbers and $\mathscr{O}$ is the ring of integers $\mathbb{Z}$, then such a lattice can always be represented by a pair $(\mathbb{Z},2m)$, i.e. by the $\mathbb{Z}$-module $\mathbb{Z}$ equipped with the $\mathbb{Z}$-bilinear form $(x,y)\mapsto 2mxy$, where m is a positive integer. The main difference here is that, for a number field K of class number greater than 1, a finitely generated, torsion free $\mathscr{O}$-module is not in general isomorphic to $\mathscr{O}$, but only to a fractional $\mathscr{O}$-ideal, whose ideal class might be not trivial.

A Jacobi form over K of half integral weight k and index $\underline{L} = (\mathfrak{c},\omega)$ is a holomorphic function $\phi(\tau,z)$ on $\mathbb{H}^n\times\mathbb{C}^n$ such that the function $\psi(\tau,x,y) := \phi(\tau,x\tau+y)\,e^{2\pi i\,\mathrm{tr}(\frac{1}{2}M(\omega)x^2\tau)}$ satisfies:

(i) The function $\psi(\tau, x, y)$ is quasi-periodic in the variables x and y in $\mathbb{R}^n$ with respect to the $\mathcal{O}$-sublattice $M(\mathfrak{c})$.
(ii) For fixed x and y in $M(K)$, the map $\tau \mapsto \psi(\tau, x, y)$ defines a Hilbert modular form of weight k (possibly with character) on the principal congruence subgroup $\Gamma(\mathfrak{a})$ of $\mathrm{SL}(2, \mathcal{O})$, where $\mathfrak{a}$ is the square of the least common multiple of the denominators of $a\mathfrak{c}^{-1}$ and of $b\mathfrak{c}^{-1}$ with $x = M(a)$ and $y = M(b)$.

Here M denotes the Minkowski embedding of K into $\mathbb{R}^n$, which maps a to the vector whose j-th component equals $\sigma_j(a)$, where we use a fixed enumeration σ_1, $\ldots$, σ_n of the embeddings of K into $\mathbb{R}$. Moreover, when writing $x\tau + y$ or $M(\omega)x^2\tau$, we view $\mathbb{C}^n$ as a ring with respect to component-wise multiplication. Finally, $\mathrm{tr}(z)$, for z in $\mathbb{C}^n$, denotes the sum of the components of z.

Note that the first property expresses the fact that, for fixed τ, the map $z \mapsto \phi(\tau, z)$ defines a theta function (a holomorphic section in a line bundle) on the abelian variety $\mathbb{C}^n / \big(M(\mathcal{O})\tau + M(\mathcal{O})\big)$. For a precise definition of Jacobi forms over number fields we refer the reader to Definition 3.45. A justification of the informal description given here can be found in the Appendix of Chap. 3. Later, it will also be more convenient to use $\mathbb{C} \otimes_{\mathbb{Q}} K$ instead of $\mathbb{C}^n$ since the first object carries naturally several algebraic structures which we shall make use of, and it allows for coordinate independent calculations.

One of the first steps into an interesting theory of Jacobi forms is, of course, to exhibit explicit examples. As it turns out, for number fields different from $\mathbb{Q}$, it is in fact already not trivial and challenging to construct examples. In this monograph, after developing a sufficiently general theory of Jacobi forms over number fields, we determine explicitly all *singular Jacobi forms over number fields*, i.e. all Jacobi forms over number fields whose weight equals $1/2$ (see Definition 3.47 and Proposition 4.1).

The singular Jacobi forms over $\mathbb{Q}$ have been determined by Skoruppa in [Sko85, p. 27]. Namely, for $\tau \in \mathbb{H}$ and $z \in \mathbb{C}$, set

$$\vartheta(\tau, z) := \sum_{s \in \mathbb{Z}} \left(\tfrac{-4}{s}\right) q^{s^2/8} \zeta^{s/2} \quad \left(q^n(\tau) := e^{2\pi i n\tau}, \zeta^n(z) := e^{2\pi i s z}\right).$$

(Here $\left(\frac{-4}{s}\right)$ denotes the nontrivial Dirichlet character modulo 4). The function ϑ is a Jacobi form over $\mathbb{Q}$ on the full modular group of weight $1/2$ and index $1/2$. In particular, ϑ is a singular Jacobi form. Skoruppa [Sko85, p. 27] showed that $\vartheta(\tau, dz)$ and $\vartheta^*(\tau, dz)$, where

$$\vartheta^*(\tau, z) := \frac{\vartheta(\tau, 2z)}{\vartheta(\tau, z)} \eta(z) = \sum_{s \in \mathbb{Z}} \left(\tfrac{12}{s}\right) q^{s^2/24} \zeta^{s/2},$$

and where d is a positive integer, are the only singular Jacobi forms over $\mathbb{Q}$ on the full modular group.

What makes the singular Jacobi forms interesting is that they occur in various important areas of mathematics. First of all, $\vartheta(\tau, z)$ is, up to normalization, the

Weierstrass' sigma-function $\sigma(\tau, z)$ associated with the elliptic curve $\mathbb{C}/(\mathbb{Z}\tau + \mathbb{Z})$. Namely, we have

$$\vartheta(\tau, z) = \eta(\tau)^3 e^{z^2 q \frac{d}{dq} \log \eta(\tau)} \sigma(\tau, z),$$

where $\eta(\tau) = q^{1/24} \prod_{n \geq 1}(1 - q^n)$ is the Dedekind's eta function. As such $\vartheta(\tau, z)$ is the basic functions out of which can be constructed all theta functions on elliptic curves. In the arithmetic theory of elliptic curves, it shows up as the Green's function for the elliptic curve $\mathbb{C}/(\mathbb{Z}\tau + \mathbb{Z})$. Moreover, $\vartheta(\tau, z)$ and $\vartheta^*(\tau, z)$ show up in the theory of Kac–Moody algebras via the famous triple and quintuple identity, respectively. For example, the Jacobi triple product identity

$$\vartheta(\tau, z) = q^{1/8}(\zeta^{1/2} - \zeta^{-1/2}) \prod_{n \geq 1}(1 - q^n)(1 - q^n\zeta)(1 - q^n\zeta^{-1})$$

can be interpreted as the Weyl–Kac denominator identity for a certain affine Kac–Moody algebra.

In view of the indicated importance of the function $\vartheta(\tau, z)$, it is natural to ask whether such functions exist also for the abelian varieties $\mathbb{C}^n/(M(\mathcal{O})\tau + M(\mathcal{O}))$ mentioned above. It is then also natural to expect that they are also singular Jacobi forms, which explains our interest in determining all singular Jacobi forms over number fields.

We explain our main results concerning singular Jacobi forms (see Theorems 4.2 and 4.3 for more precise statements).

Theorem *There exist nonzero singular Jacobi forms over K if and only if* 2 *splits completely in K and the principal genus of K contains an ideal of the form $\mathfrak{g}\mathfrak{d}^{-1}$, where $\mathfrak{g}$ is a (possibly empty) product of pairwise different prime ideals of degree* 1 *over* 3, *and where $\mathfrak{d}$ denotes the different of K.*

Recall that the principal genus of K is the set of fractional $\mathcal{O}$-ideals $\mathfrak{a}$ which represent a square in the narrow ideal class group $\mathrm{Cl}^+(K)$ of K, i.e. for which there exist a fractional $\mathcal{O}$-ideal $\mathfrak{c}$ and a totally positive ω in K such that $\mathfrak{a} = \mathfrak{c}^2\omega$. A theorem of Hecke [Hec81, Theorem 177] states that the different $\mathfrak{d}$ is a square in the ideal class group of K. However, it need not necessarily to be a square in the narrow ideal class group. A counterexample is provided by the number field $\mathbb{Q}(\sqrt{47})$. Note that 2 splits completely in this number field.

Theorem *Suppose* 2 *splits completely in K. If $\mathfrak{c}$ is a fractional $\mathcal{O}$-ideal and ω is a totally positive element in K such that $\mathfrak{g} := \mathfrak{c}^2\omega\mathfrak{d}$ is a (possibly empty) product of pairwise different prime ideals of degree* 1 *over* 3, *then*

$$\vartheta_{(\mathfrak{c},\omega)}(\tau, z) := \sum_{s \in \mathfrak{c}\mathfrak{g}^{-1}} \chi_{4\mathfrak{g}}(s') e^{2\pi i \operatorname{tr}\left(\frac{1}{8}M(\omega s^2)\tau\right)} e^{2\pi i \operatorname{tr}\left(\frac{1}{2}M(\omega s)z\right)}$$

defines a Jacobi form over K of singular weight $1/2$ and index $(\mathfrak{c}, \omega)$. Here $s' \in \mathcal{O}$ is such that $s \equiv s'\gamma \bmod 4\mathfrak{c}$, where $\gamma + 4\mathfrak{c}$ is a generator for $\mathfrak{c}\mathfrak{g}^{-1}/4\mathfrak{c}$. By $\chi_{4\mathfrak{g}}$, we denote the totally odd Dirichlet character modulo $4\mathfrak{g}$ (see Definition 2.44). Vice versa, every nonzero singular Jacobi form over K is (up to multiplication by a constant) of this form.

If $(\mathfrak{c}, \omega)$ is an index as in the theorem, then $(a^{-1}\mathfrak{c}, a^2\omega)$, for any nonzero a in K, is also such an index. Two indices are isomorphic if and only if one can be obtained from the other in this way, i.e. by multiplying with a suitable a (see Proposition 3.9). Note that the singular Jacobi forms associated with isomorphic lattices differ only in a trivial way. Namely, we have $\vartheta_{(a^{-1}\mathfrak{c}, a^2\omega)}(\tau, z) = \vartheta_{(\mathfrak{c},\omega)}(\tau, M(a)z)$. We shall see (Proposition 4.7) that the number of indices modulo isomorphism which admits a nonzero singular Jacobi form equals $|\operatorname{F}(K)| \cdot |\operatorname{Cl}^+(K)[2]|$, where $\operatorname{F}(K)$ is the subset of the principal genus consisting of ideals of the form $\mathfrak{g}\mathfrak{d}^{-1}$ with $\mathfrak{g}$ as in the last theorem, and where $\operatorname{Cl}^+(K)[2]$ is the kernel of the squaring map of the narrow ideal class group. For the field of rational numbers this number equals 2. The two classes of indices admitting a nonzero Jacobi form are represented by $(\mathbb{Z}, 1)$ and $(\mathbb{Z}, 3)$ and, indeed, we rediscover the forms from Skoruppa's theorem: $\vartheta_{(\mathbb{Z},1)} = \vartheta$ and $\vartheta_{(\mathbb{Z},3)} = \vartheta^*$.

We explain the other main themes of the book. In Chap. 3 we shall develop a general theory of Jacobi forms over number fields whose indices are arbitrary $\mathcal{O}$-lattices. In Chap. 4, we shall see that singular Jacobi forms correspond to one-dimensional submodules of certain (projective) $\operatorname{SL}(2, \mathcal{O})$-modules of theta functions (see Proposition 4.1) which turn out to be isomorphic to Weil representations associated with certain finite quadratic modules over number fields. A theory of finite quadratic modules over number fields and a theory of Weil representations associated with finite quadratic modules over number fields have not yet been worked out in the literature. Therefore we shall develop these theories in Chaps. 1 and 2, respectively. In Chap. 2, we decompose, in particular, the spaces of cyclic Weil representations into irreducible subrepresentations (see Theorem 2.5). This will give us the clue for determining explicitly all singular Jacobi forms whose indices are $\mathcal{O}$-lattices of rank one, since these correspond to the one-dimensional subrepresentations of cyclic Weil representations (see Theorem 2.6). Translating these results back to the language of Jacobi forms, we can then determine in Chap. 4 explicitly all singular Jacobi forms whose indices are $\mathcal{O}$-lattices of rank one. In the last section of this chapter we show how to construct explicitly Jacobi forms over number fields of non-singular weight, and we give examples. Finally, in the Appendix, we present tables concerning the first number fields which admit nonzero singular Jacobi forms.

Notations

In general, if the number field K is clear from the context, we often drop the subscript K, i.e. we write $\mathcal{O}$, $\mathfrak{d}$, $\mathrm{tr}(a)$, $\mathrm{N}(a)$, etc. for $\mathcal{O}_K$, $\mathfrak{d}_K$, $\mathrm{tr}_{K/\mathbb{Q}}(a)$, and $\mathrm{N}_{K/\mathbb{Q}}(a)$.

Let K be number field with ring of integers $\mathcal{O}$. A Dirichlet character modulo an integral $\mathcal{O}$-ideal $\mathfrak{a}$ is a map χ from $\mathcal{O}$ to $\mathbb{C}^*$ defined by

$$\chi(r) = \begin{cases} \chi'(r + \mathfrak{a}) & \text{if } (r, \mathfrak{a}) = 1 \\ 0 & \text{otherwise,} \end{cases}$$

where χ' is a group homomorphism from $(\mathcal{O}/\mathfrak{a})^*$ to $\mathbb{C}^*$.

An exact divisor $\mathfrak{b}$ of an integral $\mathcal{O}$-ideal $\mathfrak{a}$ is the ideal so that $\mathfrak{b} + \mathfrak{a}\mathfrak{b}^{-1} = \mathcal{O}$.

Let $\mathfrak{a}$ be a fractional $\mathcal{O}$-ideal and $\mathfrak{p}$ be a prime ideal of the number field K. We use $v_{\mathfrak{p}}(\mathfrak{a})$ for the valuation of $\mathfrak{a}$ at $\mathfrak{p}$, i.e. for the exponent of the exact power of $\mathfrak{p}$ occurring in the prime ideal factorization of $\mathfrak{a}$. Note that $v_{\mathfrak{p}}(\mathfrak{a})$ can be negative for some $\mathfrak{a}$. If $\mathfrak{a}$ is integral, we have $v_{\mathfrak{p}}(\mathfrak{a}) \geq 0$, for all $\mathfrak{p}$.

In expressions like $\sum_{\mathfrak{b}|\mathfrak{a}} \cdots$, where $\mathfrak{a}$ is an integral $\mathcal{O}$-ideal, it is always understood, if not otherwise stated, that $\mathfrak{b}$ runs through the integral $\mathcal{O}$-ideals dividing $\mathfrak{a}$. Similarly, in expressions like $\prod_{\mathfrak{p}|\mathfrak{a}} \cdots$ or $\prod_{\mathfrak{p}^a \| \mathfrak{a}} \cdots$ it is understood that $\mathfrak{p}$ runs through the prime ideals or exact prime ideal powers $\mathfrak{p}^a$ dividing $\mathfrak{a}$.

For a finite set M, the symbol $\mathbb{C}[M]$ stands for the $\mathbb{C}$-vector space of all functions from M into $\mathbb{C}$. A basis for this vector space is the set of all functions e_x $(x \in M)$ such that $e_x(y)$ equals 1 or 0 accordingly $x = y$ or not.

Let H be a subgroup of finite index in the group G, and let ϕ be a complex-valued function on G which takes the same value on each coset of G/H. We use $\sum_{g \in G/H} \phi(g)$ as a short-hand notation for $\sum_{g \in R} \phi(g)$, where R is a complete set of representatives for G/H.

In the sequel theorems are numbered independently, whereas the numbering of lemmas, propositions, examples, and corollaries share the same numbering sequence.

Chapter 1
Finite Quadratic Modules

Let K be a number field of degree n over $\mathbb{Q}$. We shall use $\mathcal{O}$, and $\mathfrak{d}$ for the ring of integers and for the different of K, respectively.

In this chapter we shall develop a theory of finite quadratic modules over number fields, i.e. a theory of finite $\mathcal{O}$-modules equipped with a quadratic form $\mathcal{O} \to K/\mathfrak{d}^{-1}$. A special emphasis is on cyclic finite quadratic modules. The main results of this chapter are Theorems 1.1 and 1.2 which give normal forms for cyclic finite quadratic modules and describe explicitly their isotropic submodules and the corresponding quotients. These results will be used in the next chapter when we decompose the spaces of cyclic Weil representations. In Sect. 1.1, we shall give the definition of finite quadratic $\mathcal{O}$-modules, and we discuss their basic properties. In Sect. 1.2, we shall specialize to cyclic $\mathcal{O}$-modules, we shall prove the two mentioned theorems and we study the orthogonal groups of cyclic finite quadratic modules which will be very crucial for the splittings of the spaces of cyclic Weil representations. Finally, in Sect. 1.3, we shall provide some lemmas concerning the quotient rings $\mathcal{O}/\mathfrak{a}$ of $\mathcal{O}$ modulo an integral $\mathcal{O}$-ideal $\mathfrak{a}$ which we shall need in Sect. 2.6 of Chap. 2.

1.1 Finite Quadratic $\mathcal{O}$-Modules

In this section we shall develop a basic theory of finite quadratic $\mathcal{O}$-modules. We shall follow closely [Sko10, Sect. 1.1], where such a theory was developed for $K = \mathbb{Q}$.

Definition 1.1 A *finite quadratic $\mathcal{O}$-module*, in short $\mathcal{O}$-FQM, is a pair (M, Q), where M is a finite $\mathcal{O}$-module, and where Q is a *non-degenerate quadratic form on M*, i.e. where $Q : M \to K/\mathfrak{d}^{-1}$ is a map which satisfies the following properties:

H. Boylan, *Jacobi Forms, Finite Quadratic Modules and Weil Representations over Number Fields*, Lecture Notes in Mathematics 2130,
DOI 10.1007/978-3-319-12916-7_1

(i) For all $a \in \mathcal{O}$ and $x \in M$ one has $Q(ax) = a^2 Q(x)$.
(ii) The map $B : M \times M \to K/\mathfrak{d}^{-1}$ defined by $B(x, y) := Q(x + y) - Q(x) - Q(y)$ is $\mathcal{O}$-bilinear and symmetric.
(iii) B is non-degenerate, i.e. $B(x, M) = \{0\}$ if and only if $x = 0$.

Let $\underline{M} = (M, Q)$ and $\underline{N} = (N, R)$ be $\mathcal{O}$-FQM. We say that there is an *isomorphism* between $\underline{M}$ and $\underline{N}$, in symbols $\underline{M} \simeq \underline{N}$, if there exists an $\mathcal{O}$-module isomorphism $\varphi : M \to N$ such that $R \circ \varphi = Q$. Two $\mathcal{O}$-FQM are called *isomorphic* if there is an isomorphism between them. The automorphisms of a finite quadratic module, i.e. the isomorphisms $\underline{M} \to \underline{M}$, form a group with respect to the composition of maps, which we denote by $\mathrm{O}(\underline{M})$ and which we call the *orthogonal group of* $\underline{M}$.

In the sequel, when we write $x \in \underline{M}$, we mean that x is an element of M, and we write $U \subseteq \underline{M}$ if U is a subset of M. Moreover, we refer to an $\mathcal{O}$-submodule U of M simply as a *submodule of* $\underline{M}$.

Example 1.2 Let $\underline{L} = (L, \beta)$ be an even $\mathcal{O}$-lattice, i.e. let L be a finitely generated torsion-free $\mathcal{O}$-module and let β be a symmetric non-degenerate $\mathcal{O}$-bilinear form on L taking values in $\mathfrak{d}^{-1}$ such that $\beta(x, x) \in 2\mathfrak{d}^{-1}$ for all $x \in L$ (see Sect. 3.1 for a short resumé of the notion of $\mathcal{O}$-lattices). The discriminant module of $\underline{L}$ is the $\mathcal{O}$-FQM

$$D_{\underline{L}} = \left(L^\#/L, x + L \mapsto \frac{1}{2}\beta(x, x) + \mathfrak{d}^{-1}\right).$$

Here $L^\#$ stands for the dual lattice of L, i.e. the set of all $y \in K \otimes_{\mathcal{O}} L$ such that $\beta(y, L) \subseteq \mathfrak{d}^{-1}$.

We have a map $\mathrm{Tr} : K/\mathfrak{d}^{-1} \to \mathbb{Q}/\mathbb{Z}, a + \mathfrak{d}^{-1} \mapsto \mathrm{tr}\,(a) + \mathbb{Z}$. It is easy to see that this map is well-defined. Indeed, if $b \in a + \mathfrak{d}^{-1}$, say, $b = a + t$ for some $t \in \mathfrak{d}^{-1}$, then $\mathrm{Tr}(t)$ is in $\mathbb{Z}$, hence, we have $\mathrm{tr}(a) \equiv \mathrm{tr}(b) \bmod \mathbb{Z}$.

Proposition 1.3 *Let* $\underline{M} = (M, Q)$ *be an* $\mathcal{O}$*-FQM. The tuple* $\mathrm{Tr}(\underline{M}) := (M, \mathrm{Tr} \circ Q)$ *defines a finite quadratic* $\mathbb{Z}$*-module.*

Proof The form $\mathrm{Tr} \circ Q$ is obviously a quadratic form on M, viewed as a $\mathbb{Z}$-module. We need to show that it is non-degenerate. Suppose $\mathrm{Tr}\left(B(x, M)\right) = \{0\}$ for some $x \in M$. Since, for all $a \in \mathcal{O}$, we have $aM \subseteq M$, we then have $\mathrm{Tr}\left(aB(x, M)\right) = \{0\}$ for all a. It is easy to see from the very definition of the different that this implies that $B(x, M) = \{0\}$. Since $\underline{M}$ is a non-degenerate $\mathcal{O}$-FQM, we conclude $x = 0$. □

Definition 1.4 Let $\underline{M} = (M, Q)$ be an $\mathcal{O}$-FQM. We call

$$\mathrm{Level}(\underline{M}) := \{a \in \mathcal{O} : aQ = 0\}$$
$$\mathrm{Ann}(\underline{M}) := \{a \in \mathcal{O} : aM = 0\}$$

the *level* and the *annihilator of* $\underline{M}$, respectively.

Note that Level($\underline{M}$) and Ann($\underline{M}$) are integral $\mathcal{O}$-ideals of K. An $\mathcal{O}$-FQM $\underline{M}$ which is annihilated by a power of a prime ideal $\mathfrak{p}$ is called, by abuse of language, a $\mathfrak{p}$-*module*. If K equals the field of rational numbers, we also call the positive integer generating Level($\underline{M}$) the level of $\underline{M}$. Moreover, in this case the positive integer generating the annihilator of $\underline{M}$ is the usual exponent of the abelian group M.

Proposition 1.5 *Let $\underline{M} = (M, Q)$ be an $\mathcal{O}$-FQM. The following holds true:*

$$\text{Level}(\underline{M}) \subseteq \text{Ann}(\underline{M}) \subseteq 1/2\,\text{Level}(\underline{M}).$$

Proof Let B be associated bilinear form of $\underline{M}$. We prove the first inclusion. Let u be in Level($\underline{M}$). So, we have $uQ = 0$. This implies that $uB = 0$, i.e $B(ux, y) = 0$ for all $x, y \in M$. From the non-degeneracy of $\underline{M}$ we conclude that $ux = 0$. Hence, $u \in \text{Ann}(\underline{M})$. Therefore, $\text{Level}(\underline{M}) \subseteq \text{Ann}(\underline{M})$.

Now we prove the second inclusion. Let $a \in \text{Ann}(\underline{M})$. Since B is $\mathcal{O}$-bilinear, $B(aM, y) = \{0\}$. In particular, $aB(x, x) = 0$ holds true for all x in M, hence we have $2aQ(x) = 0$. So $2a \in \text{Level}(\underline{M})$. Therefore, $\text{Ann}(\underline{M}) \subseteq 1/2\,\text{Level}(\underline{M})$. □

There are three operations which we can perform in the category of $\mathcal{O}$-FQM: twisting, taking direct sums and quotients. *Twisting* is the operation which maps $\underline{M} = (M, Q)$ to $\underline{M}^a := (M, aQ)$, where $a \in \mathcal{O}$ and $a \nmid \text{Level}(\underline{M})$. The latter ensures that $\underline{M}^a$ is still non-degenerate. Let (M, Q) and (N, R) be $\mathcal{O}$-FQM with associated bilinear forms B and B', respectively. We define their *direct sum* as

$$\underline{M} + \underline{N} := (M \oplus N, (x, y) \mapsto Q(x) + R(y)),$$

where $M \oplus N$ is the direct sum of the abelian groups M and N. It is clear that the map $x \oplus y \mapsto Q(x) + R(y)$ defines a non-degenerate quadratic form, so that $\underline{M} + \underline{N}$ is indeed an $\mathcal{O}$-FQM. Note that the bilinear form associated to $\underline{M} + \underline{N}$ is given by the map $(k \oplus l, x \oplus y) \mapsto B(k, x) + B'(l, y)$. Similarly, we can define the direct sum of an arbitrary finite number of $\mathcal{O}$-FQM.

An important application of taking direct sums is the decomposition of a $\mathcal{O}$-FQM into local parts. For explaining this let $\underline{M} = (M, Q)$ be a $\mathcal{O}$-FQM with associated bilinear form B. We use $\underline{M}(\mathfrak{p})$ for the $\mathcal{O}$-submodule of M which contains all elements of M that are annihilated by a power of a prime ideal $\mathfrak{p}$. The $\mathcal{O}$-FQM $\underline{M}(\mathfrak{p}) := \big(M(\mathfrak{p}), Q|_{M(\mathfrak{p})}\big)$ is called the $\mathfrak{p}$-*part of* $\underline{M}$. Note that it is a $\mathfrak{p}$-module. The non-degeneracy of the quadratic from $Q|_{M(\mathfrak{p})}$ follows from the following proposition.

Proposition 1.6 *Let $\underline{M} = (M, Q)$ be an $\mathcal{O}$-FQM. The quadratic form $Q|_{M(\mathfrak{p})}$ is non-degenerate for every prime ideal $\mathfrak{p}$. Moreover, we have*

$$\underline{M} \simeq \coprod_{\mathfrak{p} \mid \text{Ann}(\underline{M})} \underline{M}(\mathfrak{p}).$$

Proof We define

$$\varphi: \coprod_{\mathfrak{p}|\,\mathrm{Ann}(\underline{M})} \underline{M}(\mathfrak{p}) \to \underline{M}, \quad \{x_\mathfrak{p}\}_{\mathfrak{p}|\,\mathrm{Ann}(\underline{M})} \mapsto \sum_{\mathfrak{p}|\,\mathrm{Ann}(\underline{M})} x_\mathfrak{p}.$$

We show first of all that φ is surjective. Set $I_\mathfrak{p} := \prod_{\mathfrak{q}^n \| \mathrm{Ann}(\underline{M}),\, \mathfrak{q} \neq \mathfrak{p}} \mathfrak{q}^n$. Note that the ideals $I_\mathfrak{p}$ $(\mathfrak{p}|\,\mathrm{Ann}(\underline{M}))$ are relatively prime, i.e. there exists numbers $\alpha_\mathfrak{p}$ in $I_\mathfrak{p}$ such that $\sum_{\mathfrak{p}|\,\mathrm{Ann}(\underline{M})} \alpha_\mathfrak{p} = 1$. Let $x \in M$ be arbitrary. Then, $\alpha_\mathfrak{p} x$ is an element of $\underline{M}(\mathfrak{p})$ and we have $\varphi(\{\alpha_\mathfrak{p} x\}_\mathfrak{p}) = \sum_\mathfrak{p} \alpha_\mathfrak{p} x = (\sum_\mathfrak{p} \alpha_\mathfrak{p})x = x$.

We show that the quadratic form $Q|_{M(\mathfrak{p})}$ on $\underline{M}(\mathfrak{p})$ is non-degenerate for any prime ideal $\mathfrak{p}$ dividing $\mathrm{Ann}(\underline{M})$. First we shall show that for any x in $\underline{M}(\mathfrak{p})$ and y in $\underline{M}(\mathfrak{q})$, $B(x, y) = 0$, where the $\mathfrak{p}$ and the $\mathfrak{q}$ are different prime ideals. Fix $\mathfrak{p}^n \| \mathrm{Ann}(\underline{M})$ and $\mathfrak{q}^m \| \mathrm{Ann}(\underline{M})$. Since $\mathfrak{p} \neq \mathfrak{q}$, we have $\mathfrak{p}^n + \mathfrak{q}^m = \mathcal{O}$, i.e. there exists $a \in \mathfrak{p}^n$, $b \in \mathfrak{q}^m$ such that $a + b = 1$. If $x \in \underline{M}(\mathfrak{p})$ and $y \in \underline{M}(\mathfrak{q})$. we have $ax = by = 0$. Hence, $B(x, y) = (a + b)B(x, y) = B(ax, y) + B(x, by) = 0$. Suppose that we have $B(x, M(\mathfrak{p})) = \{0\}$. Using the above fact, we have $B\left(x, \coprod_{\mathfrak{q}|\,\mathrm{Ann}(\underline{M})} \underline{M}(\mathfrak{q})\right) = B(x, M) = \{0\}$. So $x = 0$, since the quadratic form Q is non-degenerate. Therefore the proposition follows. □

For explaining the third operation, namely, taking quotients, we need some preparations. Let $\underline{M} = (M, Q)$ be an $\mathcal{O}$-FQM with associated bilinear form B and U be an $\mathcal{O}$-submodule of M. The *dual group of* U is defined as:

$$U^\# := \{y \in \underline{M} : B(U, y) = 0\}. \tag{1.1}$$

Note that $U^\#$ is also an $\mathcal{O}$-submodule of M.

Proposition 1.7 *Let $\underline{M} = (M, Q)$ be an $\mathcal{O}$-FQM with associated bilinear form B and U be an $\mathcal{O}$-submodule of M. The application $x \mapsto B(x, \cdot)$ defines an exact sequence of $\mathcal{O}$-modules:*

$$0 \to U^\# \to M \to \mathrm{Hom}(U, K/\mathfrak{d}^{-1}) \to 0.$$

Here $\mathrm{Hom}(U, K/\mathfrak{d}^{-1})$ *denotes the group of $\mathcal{O}$-module homomorphisms of U into $K/\mathfrak{d}^{-1}$. In particular, one has $|U| \cdot |U^\#| = |M|$ and $(U^\#)^\# = U$.*

Proof The sequence is exact at M, since $U^\#$ is by definition the kernel of the map $M \to \mathrm{Hom}(U, K/\mathfrak{d}^{-1})$, $x \mapsto B(x, \cdot)$. The surjectivity of this map can be seen as follows: every element in $\mathrm{Hom}(U, K/\mathfrak{d}^{-1})$ can be extended to an element of $\mathrm{Hom}(M, K/\mathfrak{d}^{-1})$ [Ser73, Chap. VI,Sect. 1, Proposition 1]. The latter group has order $|M|$ [Ser73, Chap. VI,Sect. 1, Proposition 2], and it is injective (since non-degenerate) the map $x \mapsto B(x, \cdot)$ from M into $\mathrm{Hom}(M, K/\mathfrak{d}^{-1})$ is therefore also surjective. The exactness of the sequence implies that $\mathrm{Hom}(U, K/\mathfrak{d}^{-1}) \simeq M/U^\#$. Because the groups are finite and $|\mathrm{Hom}(U, K/\mathfrak{d}^{-1})| = |U|$, we obtain $|U| \cdot |U^\#| = |M|$. We have trivially $U \subseteq (U^\#)^\#$. Applying the equality for group orders to $U^\#$ instead of U, we obtain $|U||U^\#| = |M| = |U^\#||(U^\#)^\#|$, hence $U = (U^\#)^\#$. □

If U is contained in $U^\#$, then B induces a well-defined bilinear form on $U^\#/U$ as follows:

$$\underline{B} : U^\#/U \times U^\#/U \to K/\mathfrak{d}^{-1}, \quad (x + U, y + U) \mapsto B(x, y).$$

The bilinear form $\underline{B}$ is non-degenerate. Indeed, let $x + U$ be in $U^\#/U$ and suppose $\underline{B}(x + U, y + U) = 0$ for all $y \in U^\#$, i.e. $x \in (U^\#)^\#$. Proposition 1.7 implies then $x \in U$. Although the application $(x + U, y + U) \mapsto B(x, y)$ defines a bilinear form on $U^\#/U$, the application $x + U \mapsto Q(x)$ is not well-defined unless Q vanishes on U. We call an element x of the $\mathcal{O}$-FQM $\underline{M}$ *isotropic*, if $Q(x) = 0$. An $\mathcal{O}$-submodule U of M is called *isotropic*, if Q vanishes on U. If U is isotropic then U is contained in $U^\#$ and the considerations above show that the application $\underline{Q} : x + U \mapsto Q(x)$, which is now well-defined, defines a non-degenerate quadratic form on $U^\#/U$. We set

$$\underline{M}/U := \left(U^\#/U, \underline{Q}\right)$$

and call $\underline{M}/U$ the *quotient of* $\underline{M}$ by the isotropic submodule U.

Definition 1.8 Let $\underline{M} = (M, Q)$ be an $\mathcal{O}$-FQM. We define

$$\sigma(\underline{M}) := \frac{1}{\sqrt{|M|}} \sum_{x \in \underline{M}} e\{-Q(x)\}.$$

This value is called the *σ-invariant of* $\underline{M}$.

Remark Let $\underline{M}$ and $\underline{N}$ be $\mathcal{O}$-FQM. It is easy to see directly from the definition that $\sigma(\underline{M} + \underline{N}) = \sigma(\underline{M})\sigma(\underline{N})$ and $\sigma(\underline{M}^{-1}) = \overline{\sigma(\underline{M})}$.

Proposition 1.9 *Let $\underline{M} = (M, Q)$ be an $\mathcal{O}$-FQM and U be an isotropic submodule of $\underline{M}$. Then we have*

$$\sigma(\underline{M}) = \sigma(\underline{M}/U).$$

Proof Let B be associated bilinear form of $\underline{M}$. Let R denote a system of representatives for the cosets in $\underline{M}/U$. We write

$$\sigma(\underline{M}) = \sum_{x \in R} \sum_{y \in U} e\{-Q(x + y)\}.$$

Since U is isotropic we have $Q(x + y) = Q(x) + B(x, y)$. The inner sum becomes 0 unless $x \in U^\#$, when it equals $|U|$. The result is now obvious. □

Proposition 1.10 *Let $\underline{M} = (M, Q)$ be an $\mathcal{O}$-FQM. Then $\sigma(\underline{M})$ has absolute value* 1.

Proof Let B be associated bilinear form on $\underline{M}$. We write

$$|\sigma(\underline{M})|^2 = \frac{1}{|M|} \sum_{x,y \in \underline{M}} e\{-Q(x) + Q(y)\}.$$

After doing the substitution $y \mapsto y + x$ in the above sum, we obtain

$$|\sigma(\underline{M})|^2 = \frac{1}{|M|} \sum_{y \in \underline{M}} e\{Q(y)\} \sum_{x \in \underline{M}} e\{B(x, y)\}.$$

But the inner sum equals zero, unless $y = 0$, when it equals $|M|$ (see the subsequent proposition). Hence, $|\sigma(\underline{M})|^2 = 1$ as we claimed. □

Proposition 1.11 *Let $\underline{M} = (M, Q)$ be an $\mathcal{O}$-FQM with associated bilinear form B. For $y \in M$, the following holds true:*

$$S_y := \sum_{z \in \underline{M}} e\{B(z, y)\} = \begin{cases} |M| & \textit{if } y = 0 \\ 0 & \textit{otherwise.} \end{cases}$$

Proof If y equals 0, the formula is obvious. Otherwise, there exists y_0 such that $B(y_0, y) \neq 0$, since B is non-degenerate. Substituting $z \mapsto z + y_0$ we obtain that S_y equals $e\{B(y_0, y)\} S_y$. Hence, $S_y = 0$. □

1.2 Cyclic Finite Quadratic $\mathcal{O}$-Modules

In this section we shall give a full classification of cyclic finite quadratic $\mathcal{O}$-modules, their isotropic submodule, their quotients, and their orthogonal groups.

Definition 1.12 A finite quadratic $\mathcal{O}$-module (M, Q) is called a *cyclic finite quadratic $\mathcal{O}$-module*, if the $\mathcal{O}$-module M is cyclic, i.e. there exists an $x \in M$ such that $M = \mathcal{O}x$. Henceforth, a cyclic finite quadratic $\mathcal{O}$-module is called $\mathcal{O}$-CM.

Proposition 1.13 *Let $\underline{M} = (M, Q)$ be an $\mathcal{O}$-CM with level $\mathfrak{l}$. Then $(2, \mathfrak{l})^2$ divides $\mathfrak{l}$. In particular, $v_\mathfrak{p}(\mathfrak{l}) > v_\mathfrak{p}(2\mathcal{O})$ for every even prime ideal dividing $\mathfrak{l}$. For the annihilator of $\underline{M}$ we have the formula*

$$\operatorname{Ann}(\underline{M}) = \mathfrak{l}(2, \mathfrak{l})^{-1}.$$

Remark In the following we shall often tacitly use that, for integral ideals $\mathfrak{l}$ and $\mathfrak{a} = \mathfrak{l}(2, \mathfrak{l})^{-1}$, the statement $(2, \mathfrak{l})^2 | \mathfrak{l}$ is equivalent to $\mathfrak{l} | \mathfrak{a}^2$.

For the proof of the proposition we need a lemma.

Lemma 1.14 *Let $\mathfrak{a}$ be a fractional $\mathcal{O}$-ideal. The ideal $\mathfrak{b} := \sum_{a\in\mathfrak{a}} \mathcal{O}a^2$ equals $\mathfrak{a}^2$.*

Proof Multiplying by a suitable integer we can assume without loss of generality that $\mathfrak{a}$ is an integral $\mathcal{O}$-ideal. Let $a \in \mathfrak{a}$. We have $a^2 \in \mathfrak{a}^2$ and hence $\mathfrak{a}^2|\mathfrak{b}$. Vice versa, let $n = v_\mathfrak{p}(\mathfrak{b})$ for a prime ideal $\mathfrak{p}$ dividing $\mathfrak{b}$. (Recall that $v_\mathfrak{p}(\mathfrak{b})$ denotes the valuation of $\mathfrak{b}$ at $\mathfrak{p}$, see the section Notations.) There exists $a \in \mathfrak{a}$ such that $n = v_\mathfrak{p}(a^2)$. Then $n = 2k$ for some integer k. Hence $\mathfrak{p}^k \| a$. We also have that $\mathfrak{p}^{2k}|a^2$ for all $a \in \mathfrak{a}$. Hence $\mathfrak{p}^k|a$ for all $a \in \mathfrak{a}$. We therefore obtain $\mathfrak{p}^k|\mathfrak{a}$, thus, $\mathfrak{p}^n|\mathfrak{a}^2$. This proves the lemma. □

Proof of Proposition 1.13 Let B denote the bilinear form of $\underline{M}$. Write $M = \mathcal{O}\gamma$. We put $Q(\gamma) = \omega + \mathfrak{d}^{-1}$. Then we have $Q(a\gamma) = a^2\omega + \mathfrak{d}^{-1}$ for all $a \in \mathcal{O}$. First of all, we show that $\mathfrak{l}$ equals the denominator of $\omega\mathfrak{d}$. The level of $\underline{M}$ is by definition the largest $\mathcal{O}$-ideal $\mathfrak{l}$ such that $\mathfrak{l}Q = 0$. i.e. $\mathfrak{l}a^2\omega \in \mathfrak{d}^{-1}$, or, equivalently, such that $\mathfrak{l}\omega\mathfrak{d}$ is an integral $\mathcal{O}$-ideal. Hence $\mathfrak{l}$ equals the denominator of $\omega\mathfrak{d}$.

Next we prove $\mathfrak{a} := \mathrm{Ann}(\underline{M}) = \mathfrak{l}(2,\mathfrak{l})^{-1}$. By the non-degeneracy of B, the annihilator of $\underline{M}$ consists of all $a \in \mathcal{O}$ such that $B(a\gamma, a'\gamma) = 0$ for all $a' \in \mathcal{O}$. But $B(a\gamma, a'\gamma) = 2aa'\omega + \mathfrak{d}^{-1}$. Hence the annihilator of $\underline{M}$ consists of all $a \in \mathcal{O}$ such that $2a\omega\mathfrak{d}$ is integral, which is equivalent to $\mathfrak{l}(2,\mathfrak{l})^{-1}|a$. This proves the claim.

By the remark it remains to show that $\mathfrak{l}|\mathfrak{a}^2$. If $a \in \mathfrak{a}$, then $0 = Q(a\gamma) = a^2\omega + \mathfrak{d}^{-1}$. This implies that $a^2\omega \in \mathfrak{d}^{-1}$, i.e. the ideal $a^2\omega\mathfrak{d}$ is integral. Since $\mathfrak{l}$ is the denominator of $\omega\mathfrak{d}$, we have that $\mathfrak{l}|a^2$ for all $a \in \mathfrak{a}$. Since $\mathfrak{a}^2$ equals the ideal generated by the squares of elements in $\mathfrak{a}$ (see Lemma 1.14) we conclude $\mathfrak{l}|\mathfrak{a}^2$.

Finally, let $\mathfrak{p}^l$ be the exact power of an even prime dividing $\mathfrak{l}$. If $\mathfrak{p}^l$ divided 2, then $\mathfrak{l}(2,\mathfrak{l})^{-1}$ would not be divisible by $\mathfrak{p}$, in contradiction to $(2,\mathfrak{l})^2|\mathfrak{l}$. □

By the proposition the ideal $\mathfrak{l}(2,\mathfrak{l})^{-2}$, where $\mathfrak{l}$ is the level of an $\mathcal{O}$-CM, is integral. Since it will show up in many subsequent formulas we introduce a name for this quantity.

Definition 1.15 Let $\underline{M}$ be an $\mathcal{O}$-CM with level $\mathfrak{l}$. We call

$$\mathrm{Mod}(\underline{M}) := \mathfrak{l}(2,\mathfrak{l})^{-2}$$

the *modified level of $\underline{M}$*.

Theorem 1.1

(i) Let $\omega \in K^*$ and let $\mathfrak{l}$ be the denominator of $\omega\mathfrak{d}$. Assume $(2,\mathfrak{l})^2|\mathfrak{l}$. Then the pair

$$\underline{M}(\omega) := \left(\mathcal{O}/\mathfrak{a}, x + \mathfrak{a} \mapsto \omega x^2 + \mathfrak{d}^{-1}\right),$$

where $\mathfrak{a} = \mathfrak{l}(2,\mathfrak{l})^{-1}$, defines an $\mathcal{O}$-FQM with annihilator $\mathfrak{a}$ and level $\mathfrak{l}$. In fact, $\underline{M}(\omega)$ is an $\mathcal{O}$-CM with generator $1 + \mathfrak{a}$.

(ii) Every $\mathcal{O}$-CM is isomorphic to an $\mathcal{O}$-FQM of the form $\underline{M}(\omega)$.

(iii) *Two $\mathcal{O}$-CM $\underline{M}(\omega_1)$ and $\underline{M}(\omega_2)$ are isomorphic if and only if there exists an a in $\mathcal{O}$, relatively prime to $\mathfrak{l}$, such that $\omega_1 \equiv \omega_2 a^2 \bmod \mathfrak{d}^{-1}$. Here $\mathfrak{l}$ stands for the denominator of $\omega_2 \mathfrak{d}$.*

Proof First we prove (i). Note that the assumption $(2,\mathfrak{l})^2|\mathfrak{l}$ is equivalent to the statement $\mathfrak{l}|\mathfrak{a}^2$. We show that the map $Q : x + \mathfrak{a} \mapsto \omega x^2 + \mathfrak{d}^{-1}$ is well-defined and that it is non-degenerate. First note that $\omega \mathfrak{d} \mathfrak{a}^2$ is integral (by the assumption $\mathfrak{l}|\mathfrak{a}^2$). For the well-definedness, we need to have that if $y \in x + \mathfrak{a}$, then $\omega x^2 - \omega y^2 \in \mathfrak{d}^{-1}$. Write $y = x + k$ ($k \in \mathfrak{a}$). Then $\omega x^2 - \omega y^2 = -2\omega x k - \omega k^2 \in \omega(2\mathfrak{a} + \mathfrak{a}^2)$ lies in $\mathfrak{d}^{-1}$, since $\mathfrak{l}$ divides $2\mathfrak{a}$ by definition of $\mathfrak{a}$ and since $\mathfrak{l}$ divides $\mathfrak{a}^2$ by assumption. For the non-degeneracy of the quadratic form Q, we need to have that $2\omega x \mathcal{O} \subseteq \mathfrak{d}^{-1}$ ($x \in \mathcal{O}$) if and only if $x \in \mathfrak{a}$. Indeed, $2\omega x \mathfrak{d}$ is integral if and only if the denominator $\mathfrak{l}$ of $\omega\mathfrak{d}$ divides $2x$, i.e. if and only if $\mathfrak{a} = \mathfrak{l}(2,\mathfrak{l})^{-1}$ divides x. It is obvious from the construction that $\underline{M}(\omega)$ has annihilator $\mathfrak{a}$ and level $\mathfrak{l}$. The $\mathcal{O}$-FQM $\underline{M}(\omega)$ is an $\mathcal{O}$-CM, since it is generated by the multiplicative neutral element $1 + \mathfrak{a}$ of the ring $\mathcal{O}/\mathfrak{a}$.

Secondly we prove (ii). Let $\underline{M} = (M, Q)$ be a cyclic finite quadratic $\mathcal{O}$-module with annihilator $\mathfrak{a}$. Write $M = \mathcal{O}\gamma$, set $\omega = Q(\gamma)$ and let $\mathfrak{l}$ denote the denominator of $\omega\mathfrak{d}$. Then $\mathfrak{l}$ equals the level of $\underline{M}$. Indeed, since $M = \mathcal{O}\gamma$ the ideal generated by the $Q(x)$ ($x \in M$) equals $Q(\gamma)\mathcal{O}$. By Proposition 1.13 we have then $(2,\mathfrak{l})^2|\mathfrak{l}$ and $\mathfrak{a} = \mathfrak{l}(2,\mathfrak{l})^{-1}$. Hence $\underline{M}(Q(\gamma)) = (\mathcal{O}/\mathfrak{a}, x + \mathfrak{a} \mapsto Q(\gamma)x^2 + \mathfrak{d}^{-1})$. It is obvious that the map $x + \mathfrak{a} \mapsto x\gamma$ defines an isomorphism $\underline{M}(Q(\gamma)) \to \underline{M}$.

Lastly, we prove (iii). Assume that there exists an isomorphism φ from $\underline{M}(\omega_1)$ to $\underline{M}(\omega_2)$. Then the levels of both modules coincide, hence are equal to $\mathfrak{l}$. The annihilator of both $\mathcal{O}$-FQM is $\mathfrak{a} := \mathfrak{l}(2,\mathfrak{l})^{-1}$. Since $1 + \mathfrak{a}$ is a generator of the $\mathcal{O}$-module $\mathcal{O}/\mathfrak{a}$ there exist an a in $\mathcal{O}$ such that $\varphi(1 + \mathfrak{a}) = a(1 + \mathfrak{a})$. Since φ is an isomorphism there exist also an a' such that $\varphi^{-1}(1 + \mathfrak{a}) = a'(1 + \mathfrak{a})$. We conclude that $a'a \equiv 1 \bmod \mathfrak{a}$, i.e. that a is relatively prime to $\mathfrak{a}$. Since, $\mathfrak{a}$ and $\mathfrak{l}$ have the same prime divisors (see Proposition 1.13), we see that a is also relatively prime to $\mathfrak{l}$. Finally, since $\varphi(1 + \mathfrak{a}) = a + \mathfrak{a}$ and since φ preserves the quadratic forms, we find $\omega_2 a^2 \equiv \omega_1 \bmod \mathfrak{d}^{-1}$.

If, vice versa, $\omega_2 a^2 \equiv \omega_1 \bmod \mathfrak{d}^{-1}$ for some a, relatively prime to the denominator $\mathfrak{l}$ of $\omega_2\mathfrak{d}$, then $\omega_1\mathfrak{d}$ has also denominator $\mathfrak{l}$. Indeed, write $\omega_2 a^2 = \omega_1 + t$ with t in $\mathfrak{d}^{-1}$. Then $\omega_2 a^2 \mathfrak{d} \subseteq \omega_1\mathfrak{d} + \mathcal{O}$, and therefore $\mathfrak{l}_1\omega_2 a^2\mathfrak{d} \subseteq \omega_1\mathfrak{d}\mathfrak{l}_1 + \mathfrak{l}_1$, where $\mathfrak{l}_1$ denotes the denominator of $\omega_1\mathfrak{d}$, from which we deduce that $\mathfrak{l}_1\omega_2 a^2\mathfrak{d}$ is integral. Hence $\mathfrak{l}$ divides $\mathfrak{l}_1 a^2$, and since a and $\mathfrak{l}$ are coprime, we find $\mathfrak{l}|\mathfrak{l}_1$. Changing the role of $\mathfrak{l}$ and $\mathfrak{l}_1$ in the preceding argument we find also $\mathfrak{l}_1|\mathfrak{l}$. It is then clear that the map $x + \mathfrak{a} \mapsto ax + \mathfrak{a}$ defines an isomorphism of $\underline{M}(\omega_1) \to \underline{M}(\omega_2)$. □

Corollary 1.16 *The number of isomorphism classes of $\mathcal{O}$-CM with a given level $\mathfrak{l}$ equals the number of elements in $(\mathcal{O}/\mathfrak{l})^*[2]$, where $(\mathcal{O}/\mathfrak{l})^*[2]$ denotes the kernel of the squaring map of $(\mathcal{O}/\mathfrak{l})^*$.*

Remark Applying the Chinese remainder theorem [Neu99, I. 3, Theorem (3.6)] our theorem can be restated in the form that the number of isomorphism classes of $\mathcal{O}$-CM with a given level $\mathfrak{l}$ equals therefore $\prod_{\mathfrak{p}^n \| \mathfrak{l}} a(\mathfrak{p}^n)$, where $a(\mathfrak{p}^n)$ is the number of solutions of $x^2 = 1$ in $(\mathcal{O}/\mathfrak{p}^n)^*$. For odd $\mathfrak{p}$, there are exactly two solutions of $x^2 = 1$ in $(\mathcal{O}/\mathfrak{p}^n)^*$, i.e. $a(\mathfrak{p}^n) = 2$. In general, $a(\mathfrak{p}^n) = 2^e$, where e denotes the number of even elementary divisors of $(\mathcal{O}/\mathfrak{p}^n)^*$. For even $\mathfrak{p}$, the number of solutions depends very much on the arithmetic of the number field.

Proof If $\mathfrak{l}$ is given, then we can always form an $\mathcal{O}$-CM. Indeed, let $\mathfrak{b}$ be an integral $\mathcal{O}$-ideal which lie in the inverse ideal class of $\mathfrak{l}\mathfrak{d}$ which satisfies $\mathfrak{b} + \mathfrak{l} = \mathcal{O}$. Then there exists some ω in K such that $\mathfrak{b}(\mathfrak{l}\mathfrak{d})^{-1} = \mathcal{O}\omega$. Denote $\mathfrak{a} := \mathfrak{l}(2, \mathfrak{l})^{-1}$. Hence, $\big(\mathcal{O}/\mathfrak{a}, x + \mathfrak{a} \mapsto \omega x^2\big)$ defines an $\mathcal{O}$-CM (see Theorem 1.1 (i)).

Let $\underline{M}$ be a cyclic $\mathcal{O}$-module of level $\mathfrak{l}$. By Theorem 1.1 $\underline{M}$ is isomorphic to some $\underline{M}(\omega)$ where $\mathfrak{l}\omega\mathfrak{d}$ is an integral $\mathcal{O}$-ideal relatively prime to $\mathfrak{l}$, and, vice versa, for every such ω, the $\mathcal{O}$-FQM $\underline{M}(\omega)$ has level $\mathfrak{l}$. Moreover, $\underline{M}(\omega)$ depends obviously only on ω modulo $\mathfrak{d}^{-1}$. We shall prove in a moment that $\mathfrak{l}\omega\mathfrak{d}$ is an integral $\mathcal{O}$-ideal relatively prime to $\mathfrak{l}$ if and only if $\omega+\mathfrak{d}^{-1}$ generates the $\mathcal{O}$-module $\mathfrak{l}^{-1}\mathfrak{d}^{-1}/\mathfrak{d}^{-1}$. Thus if we use $\mathfrak{J}_{\mathfrak{l}}$ for the set of isomorphism classes of cyclic $\mathcal{O}$-modules of level $\mathfrak{l}$, the application $\omega \mapsto M(\omega)$ induces a surjective map from the set G of the generators $\mathfrak{l}^{-1}\mathfrak{d}^{-1}/\mathfrak{d}^{-1}$ onto $\mathfrak{J}_{\mathfrak{l}}$. By Theorem 1.1 (iii) this map induces a bijection

$$G/\big((\mathcal{O}/\mathfrak{l})^*\big)^2 \to \mathfrak{J}_{\mathfrak{l}}.$$

It is easy to see that there is an $\mathcal{O}$-module isomorphism $\mathcal{O}/\mathfrak{l} \to \mathfrak{l}^{-1}\mathfrak{d}^{-1}/\mathfrak{d}^{-1}$ (see Lemma 1.17). So, the number that we are looking for equals the number of elements in $(\mathcal{O}/\mathfrak{l})^*/((\mathcal{O}/\mathfrak{l})^*)^2$. Since we have an exact sequence

$$1 \to \mathrm{Ker}(\mathrm{Sq}) \to (\mathcal{O}/\mathfrak{l})^* \xrightarrow{\mathrm{Sq}} ((\mathcal{O}/\mathfrak{l})^*)^2 \to 1,$$

where Sq is the squaring map, we conclude that the number of elements in the group $(\mathcal{O}/\mathfrak{l})^*/((\mathcal{O}/\mathfrak{l})^*)^2$ equals the number of elements in the kernel of Sq. This proves the corollary.

It remains to determine the ω such that $\mathfrak{l}\omega\mathfrak{d}$ is an integral ideal relatively prime to $\mathfrak{l}$. Write such an ω in the form $\omega\mathfrak{d} = \mathfrak{b}\mathfrak{l}^{-1}$ with an integral $\mathfrak{b}$ coprime to $\mathfrak{l}$. Then $\omega = \mathfrak{b}\mathfrak{l}^{-1}\mathfrak{d}^{-1} \subseteq \mathcal{O}\mathfrak{l}^{-1}\mathfrak{d}^{-1} = \mathfrak{l}^{-1}\mathfrak{d}^{-1}$. Hence, $\omega \in \mathfrak{l}^{-1}\mathfrak{d}^{-1}$, and, by the assumption on ω, it is so that $\omega\mathfrak{d}\mathfrak{l} + \mathfrak{l} = \mathcal{O}$, i.e. so that $\mathcal{O}(\omega + \mathfrak{d}^{-1}) = \mathfrak{l}^{-1}\mathfrak{d}^{-1}/\mathfrak{d}^{-1}$. It is easy to see that these reasoning can be reversed. □

Our next goal is a description of all isotropic submodules of a given $\mathcal{O}$-CM and its quotients. For this we need a lemma.

Lemma 1.17 *Let $\mathfrak{a}$ and $\mathfrak{b}$ be fractional $\mathcal{O}$-ideals such that $\mathfrak{a} \subseteq \mathfrak{b}$. The quotient $\mathfrak{b}/\mathfrak{a}$ is $\mathcal{O}$-CM. Its generators are the elements $\gamma+\mathfrak{a}$, where γ is in $\mathfrak{b}$ such that $\mathfrak{b} = \mathcal{O}\gamma+\mathfrak{a}$.*

Proof Multiplying by a suitable integer, we can assume without loss of generality that $\mathfrak{a}$ and $\mathfrak{b}$ are integral $\mathcal{O}$-ideals. Let $\mathfrak{c}$ be a fractional $\mathcal{O}$-ideal in the ideal class of $\mathfrak{b}^{-1}$ which is relatively prime to the (integral) $\mathcal{O}$-ideal $\mathfrak{a}\mathfrak{b}^{-1}$. We thus have $\mathfrak{c}+\mathfrak{a}\mathfrak{b}^{-1} = \mathcal{O}$ and $\mathfrak{b}\mathfrak{c} = \gamma\mathcal{O}$, for some $\gamma \in \mathfrak{b}$. Consequently, we have $\gamma\mathcal{O} + \mathfrak{a} = \mathfrak{b}$. This implies

$\mathfrak{b}/\mathfrak{a} = \mathcal{O}(\gamma + \mathfrak{a})$. Indeed, let $b + \mathfrak{a} \in \mathfrak{b}/\mathfrak{a}$. Hence, $b = d\gamma + c$, for some $d \in \mathcal{O}$, $c \in \mathfrak{a}$. Then $b + \mathfrak{a} = d\gamma + \mathfrak{a} = d(\gamma + \mathfrak{a})$. This proves the lemma. □

Theorem 1.2 *Let $\underline{M} = (M, Q)$ be an $\mathcal{O}$-CM with level $\mathfrak{l}$, modified level $\mathfrak{m} = \mathfrak{l}\,(2, \mathfrak{l})^{-2}$ and annihilator $\mathfrak{a}$ (recall from Theorem 1.1 that $\mathfrak{a} = \mathfrak{m}\,(2, \mathfrak{l})$).*

(i) The isotropic submodules of $\underline{M}$ are of the form $\mathfrak{a}\mathfrak{b}^{-1}M$, where $\mathfrak{b}$ is an integral $\mathcal{O}$-ideal such that $\mathfrak{b}^2|\mathfrak{m}$. In particular, the sum of any two isotropic submodules is again isotropic.
(ii) If $\underline{M} = \underline{M}(\omega)$, and $\mathfrak{a}\mathfrak{b}^{-1}/\mathfrak{a}$ is an isotropic submodule of $\underline{M}(\omega)$ (so that $\mathfrak{b}^2|\mathfrak{m}$), then the quotient module $\underline{M}(\omega)/(\mathfrak{a}\mathfrak{b}^{-1}/\mathfrak{a})$ is isomorphic to the $\mathcal{O}$-FQM

$$\underline{M}(\omega\gamma^2) = \left(\mathcal{O}/\mathfrak{a}\mathfrak{b}^{-2}, x + \mathfrak{a}\mathfrak{b}^{-2} \mapsto \omega\gamma^2x^2 + \mathfrak{d}^{-1}\right),$$

where $\gamma \in \mathfrak{b}$ is such that $\mathfrak{b} = \mathcal{O}\gamma + \mathfrak{a}\mathfrak{b}^{-1}$ (see Lemma 1.17).
(iii) In particular, the class of $\mathcal{O}$-CM is closed under taking quotients.

Remark If the set of isotropic submodules of an $\mathcal{O}$-FQM is closed under taking sums then it possesses only one maximal isotropic submodule, namely the sum of all isotropic submodules. Vice versa, if it possesses only one maximal isotropic submodule then the sum of any two isotropic submodules is contained in the unique maximal one, and hence also isotropic. Thus, by part (i), an $\mathcal{O}$-CM possesses only one maximal isotropic submodule.

Example 1.18 Note that there are also $\mathcal{O}$-FQM which have this property, but are not cyclic. The $\mathcal{O}$-FQM $(\mathbb{Z}/4\mathbb{Z} \times \mathbb{Z}/4\mathbb{Z}, Q)$ with $Q(x + 4\mathbb{Z}, y + 4\mathbb{Z}) = (x^2 + xy + y^2)/4 + \mathbb{Z}$ is such an example (for $K = \mathbb{Q}$). It possesses exactly five isotropic submodules, namely, the submodules 0, $\langle(0, [2])\rangle$, $\langle([2], [2])\rangle$, $\langle([2], 0)\rangle$ and $\langle(0, [2])\rangle \oplus \langle([2], 0)\rangle$, where the last one is the unique maximal one. (Here we use $[x] = x + 4\mathbb{Z}$.)

Proof of Theorem 1.2 First we prove (i). We have that $\underline{M}$ is isomorphic to an $\mathcal{O}$-FQM $\underline{M}(\omega)$ for some nonzero $\omega \in K$ as given in Theorem 1.1. Clearly, any $\mathcal{O}$-submodule of $\underline{M}(\omega)$ is of the form $\mathfrak{c}/\mathfrak{a}$ for some integral $\mathcal{O}$-ideal $\mathfrak{c}$ such that $\mathfrak{a} \subseteq \mathfrak{c}$. Let $\mathfrak{c}/\mathfrak{a}$ be an isotropic submodule of $\underline{M}(\omega)$. For all $x \in \mathfrak{c}$, we have $\omega x^2 \in \mathfrak{d}^{-1}$ i.e. the ideal $\omega\mathfrak{d}\mathfrak{c}^2$ is integral (see Lemma 1.14). Hence, $\mathfrak{l}$ divides $\mathfrak{c}^2$. Therefore, any isotropic submodule of $\underline{M}(\omega)$ is of the form $\mathfrak{c}/\mathfrak{a}$ with some integral $\mathcal{O}$-ideal $\mathfrak{c}$ such that $\mathfrak{l}|\mathfrak{c}^2|\mathfrak{a}^2$. It is then clear that the isotropic submodules of $\underline{M}$ are of the form $\mathfrak{c}M$, where $\mathfrak{c}$ runs through the set of integral $\mathcal{O}$-ideals which satisfy $\mathfrak{l}|\mathfrak{c}^2|\mathfrak{a}^2$. However, it is easily checked that the following map is an isomorphism:

$$\{\mathfrak{c} \subseteq \mathcal{O} : \mathfrak{l}|\mathfrak{c}^2|\mathfrak{a}^2\} \to \{\mathfrak{b} \subseteq \mathcal{O} : \mathfrak{b}^2|\mathfrak{m}\}, \quad \mathfrak{c} \mapsto \mathfrak{a}\mathfrak{c}^{-1} =: \mathfrak{b}.$$

Hence, the first statement of (i) is proved.

Let U and V be two isotropic submodules of $\underline{M}$, say, $U = \mathfrak{a}\mathfrak{b}_1^{-1}M$ and $V = \mathfrak{a}\mathfrak{b}_2^{-1}M$. Then

$$U + V = \mathfrak{a}\mathfrak{b}_1^{-1}M + \mathfrak{a}\mathfrak{b}_2^{-1}M = \left(\mathfrak{a}\mathfrak{b}_1^{-1} + \mathfrak{a}\mathfrak{b}_2^{-1}\right)M = \mathfrak{a}(\mathfrak{b}_1, \mathfrak{b}_2)\mathfrak{b}_1^{-1}\mathfrak{b}_2^{-1}M,$$

and since $\mathfrak{b}_1^2$ and $\mathfrak{b}_2^2$ divide $\mathfrak{m}$ it is clear that the square of the least common multiple $\mathfrak{b}_1\mathfrak{b}_2(\mathfrak{b}_1, \mathfrak{b}_2)^{-1}$ also divides $\mathfrak{m}$, i.e. that $U + V$ is isotropic.

We prove the statement (ii). Let $\mathfrak{a}\mathfrak{b}^{-1}/\mathfrak{a}$ be an isotropic submodule of $\underline{M}(\omega)$. To determine the quotient module $\underline{M}(\omega)/(\mathfrak{a}\mathfrak{b}^{-1}/\mathfrak{a})$, we need to determine the dual module $(\mathfrak{a}\mathfrak{b}^{-1}/\mathfrak{a})^{\#}$. Using the Definition (1.1), we obtain

$$\begin{aligned}(\mathfrak{a}\mathfrak{b}^{-1}/\mathfrak{a})^{\#} &= \{x + \mathfrak{a} \in \mathcal{O}/\mathfrak{a} : 2\omega x\mathfrak{a}\mathfrak{b}^{-1} \subseteq \mathfrak{d}^{-1}\} \\ &= \{x + \mathfrak{a} \in \mathcal{O}/\mathfrak{a} : \mathfrak{b}|x\} = \mathfrak{b}/\mathfrak{a}.\end{aligned}$$

For the second equality we write $\omega\mathfrak{d} = \mathfrak{b}'\mathfrak{l}^{-1}$, where $\mathfrak{b}'$ in an integral $\mathcal{O}$-ideal such that $(\mathfrak{b}', \mathfrak{l}) = 1$. Let $x + \mathfrak{a} \in \mathcal{O}/\mathfrak{a}$. Then we have that $2\omega\mathfrak{d}x\mathfrak{a}\mathfrak{b}^{-1}$ is integral if and only if $\mathfrak{b}|x$. Indeed, since $\mathfrak{a} = \mathfrak{l}(2, \mathfrak{l})^{-1}$ (see Proposition 1.13), we have $2\omega\mathfrak{d}x\mathfrak{a}\mathfrak{b}^{-1} = 2(2, \mathfrak{l})^{-1}\mathfrak{b}'\mathfrak{b}^{-1}x$. But $\mathfrak{b}$ is relatively prime to $2(2, \mathfrak{l})^{-1}\mathfrak{b}'$, since $\mathfrak{b}$ is a divisor of $\mathfrak{l}(2, \mathfrak{l})^{-1}$. Therefore, we have

$$(\mathfrak{a}\mathfrak{b}^{-1}/\mathfrak{a})^{\#}/(\mathfrak{a}\mathfrak{b}^{-1}/\mathfrak{a}) = (\mathfrak{b}/\mathfrak{a})/(\mathfrak{a}\mathfrak{b}^{-1}/\mathfrak{a}) \simeq \mathfrak{b}/\mathfrak{a}\mathfrak{b}^{-1},$$

and hence

$$\underline{M}(\omega)/(\mathfrak{a}\mathfrak{b}^{-1}/\mathfrak{a}) \simeq \left(\mathfrak{b}/\mathfrak{a}\mathfrak{b}^{-1}, x + \mathfrak{a}\mathfrak{b}^{-1} \mapsto \omega x^2 + \mathfrak{d}^{-1}\right) =: \underline{N}.$$

Note that the annihilator of the $\mathcal{O}$-module $\mathfrak{b}/\mathfrak{a}\mathfrak{b}^{-1}$ equals the ideal $\mathfrak{a}\mathfrak{b}^{-2}$ (which is integral, since $\mathfrak{b}^2$ divides $\mathfrak{m}$ and $\mathfrak{m}$ divides $\mathfrak{a}$). By Lemma 1.17 we know that $\mathfrak{b}/\mathfrak{a}\mathfrak{b}^{-1}$ is an $\mathcal{O}$-CM, i.e. there exists $\gamma \in \mathfrak{b}$ such that $\mathfrak{b} = \mathcal{O}\gamma + \mathfrak{a}\mathfrak{b}^{-1}$. The application $x + \mathfrak{a}\mathfrak{b}^{-2} \mapsto x\gamma + \mathfrak{a}\mathfrak{b}^{-1}$ defines therefore an isomorphism

$$\left(\mathcal{O}/\mathfrak{a}\mathfrak{b}^{-2}, x + \mathfrak{a}\mathfrak{b}^{-2} \mapsto \omega\gamma^2x^2 + \mathfrak{d}^{-1}\right) \xrightarrow{\simeq} \underline{N},$$

which proves (ii).

Lastly, the statement (iii) is an immediate consequence of (i) and (ii). □

Corollary 1.19 *Let $\underline{M} = (M, Q)$ an $\mathcal{O}$-CM, and let $\mathfrak{a}$, $\mathfrak{l}$, $\mathfrak{m}$ denote its annihilator, level and modified level. Then the annihilator, the level and the modified level of the quotient module $\underline{M}/\mathfrak{a}\mathfrak{b}^{-1}M$ equals $\mathfrak{a}\mathfrak{b}^{-2}$, $\mathfrak{l}\mathfrak{b}^{-2}$ and $\mathfrak{m}\mathfrak{b}^{-2}$, respectively.*

Proof Set $U = \mathfrak{a}\mathfrak{b}^{-1}M$. By Theorem 1.2 (ii), $\underline{M}/U$ is isomorphic to some $\mathcal{O}$-FQM $\underline{M}(\omega\gamma^2)$ with $\mathfrak{b} = \mathcal{O}\gamma + \mathfrak{a}\mathfrak{b}^{-1}$ ($\gamma \in \mathfrak{b}$). Clearly, the latter has annihilator $\mathfrak{a}\mathfrak{b}^{-2}$. It is enough to show that the $\mathcal{O}$-FQM $\underline{M}(\omega\gamma^2)$ has level $\mathfrak{l}\mathfrak{b}^{-2}$. Because, then, it is clear that the modified level of $\underline{M}/U$ equals $\mathfrak{m}\mathfrak{b}^{-2}$. Write $\mathcal{O}\gamma = \mathfrak{b}\mathfrak{b}_0$ with some integral $\mathcal{O}$-ideal $\mathfrak{b}_0$. The level of $\underline{M}(\omega\gamma^2)$ equals the denominator of $\omega\mathfrak{d}\gamma^2$. To show that $\underline{M}(\omega\gamma^2)$ has level $\mathfrak{l}\mathfrak{b}^{-2}$, it is enough to show that $\mathfrak{l}\mathfrak{b}^{-2}$ is relatively prime to $\mathfrak{b}_0$, since the denominator of $\omega\mathfrak{d}$ equals $\mathfrak{l}$. The identity $\mathfrak{b} = \mathcal{O}\gamma + \mathfrak{a}\mathfrak{b}^{-1}$ implies that $(\mathfrak{b}_0, \mathfrak{a}\mathfrak{b}^{-2}) = 1$. Since $\mathfrak{b}^2|\mathfrak{m}$, we have that $(2, \mathfrak{l})|\mathfrak{a}\mathfrak{b}^{-2}$. Here we used the fact that $\mathfrak{a} = \mathfrak{m}(2, \mathfrak{l})$ (see Proposition 1.13). Hence, $(2, \mathfrak{l})$ is also relatively prime to $\mathfrak{b}_0$. Since we have $\mathfrak{l}\mathfrak{b}^{-2} = \mathfrak{a}\mathfrak{b}^{-2}(2, \mathfrak{l})$, obviously $\mathfrak{b}_0$ is relatively prime to $\mathfrak{l}\mathfrak{b}^{-2}$. This proves the corollary. □

We finally describe the orthogonal groups of $\mathcal{O}$-CM. It will turn out that $\mathrm{O}(\underline{M})$ is isomorphic to a certain subgroup of $(\mathcal{O}/\mathfrak{a})^*$, for which we introduce a special name.

Definition 1.20 Let $\underline{M} = (M, Q)$ be an $\mathcal{O}$-CM with level $\mathfrak{l}$ and annihilator $\mathfrak{a}$. We define:

$$\mathrm{E}(\underline{M}) := \{\varepsilon + \mathfrak{a} \in (\mathcal{O}/\mathfrak{a})^* : \varepsilon^2 \equiv 1 \bmod \mathfrak{l}\}.$$

Remark Note that $\mathrm{E}(\underline{M})$ is well-defined. Namely, assume $\varepsilon \equiv \varepsilon' \bmod \mathfrak{a}$. Since $(2, \mathfrak{l})^2 | \mathfrak{l} = \mathfrak{a}(2, \mathfrak{l})$ (Proposition 1.13) we deduce $(2, \mathfrak{l})|\mathfrak{a}$, hence $\varepsilon \equiv \varepsilon' \bmod (2, \mathfrak{l})$, and then $\mathfrak{a}(2, \mathfrak{l})|(\varepsilon - \varepsilon')(\varepsilon + \varepsilon')$, i.e. $\varepsilon^2 \equiv \varepsilon'^2 \bmod \mathfrak{l}$. In fact, $\mathrm{E}(\underline{M})$ is a subgroup of $(\mathcal{O}/\mathfrak{a})^*$.

Proposition 1.21 *Let $\underline{M} = (M, Q)$ be an $\mathcal{O}$-CM. The application $g \mapsto m_g$, where m_g denotes multiplication by g, defines an isomorphism of groups $\mathrm{E}(\underline{M}) \xrightarrow{\simeq} \mathrm{O}(\underline{M})$.*

Proof For the well-definedness we need to show that multiplication by an element $g = \varepsilon + \mathfrak{a} \in \mathrm{E}(\underline{M})$ defines an orthogonal transformation of $\underline{M}$. Since $\varepsilon^2 \equiv 1 \bmod \mathfrak{l}$ for any $x \in M$, we have $Q(\varepsilon x) - Q(x) = Q(x)(\varepsilon^2 - 1) = 0$, i.e. we have $Q(\varepsilon x) = Q(x)$.

The injectivity is obvious. For the surjectivity we need to show that any α in $\mathrm{O}(\underline{M})$ is given by $\alpha(x) = gx$ for some $g \in \mathrm{E}(\underline{M})$. Write $M = \mathcal{O}\gamma$ for some $\gamma \in M$ and $\alpha(\gamma) = \varepsilon\gamma$ for some $\varepsilon \in \mathcal{O}$ and $\varepsilon \notin \mathfrak{a}$. Write $x = a\gamma$ with $a \in \mathcal{O}$ and $a \notin \mathfrak{a}$. Since α is an $\mathcal{O}$-module homomorphism, we have

$$\alpha(x) = a\alpha(\gamma) = a\varepsilon\gamma = a\gamma\varepsilon = x\varepsilon.$$

This proves the proposition. □

Proposition 1.22 *Let $\underline{M}$ be an $\mathcal{O}$-CM with annihilator $\mathfrak{a}$ and modified level $\mathfrak{m}$. Then the map $\varepsilon + \mathfrak{a} \mapsto \{\varepsilon + \mathfrak{p}^a\}_\mathfrak{p}$ defines an isomorphism of groups $\mathrm{E}(\underline{M}) \simeq \coprod_{\mathfrak{p}^a \| \mathfrak{a}} \mathrm{E}(\underline{M}(\mathfrak{p}))$. Via this isomorphism the factor $\mathrm{E}(\underline{M}(\mathfrak{p}))$ corresponds to the subgroup $\langle \varepsilon_\mathfrak{p} + \mathfrak{a} \rangle$ of $\mathrm{E}(\underline{M})$, where $\varepsilon_\mathfrak{p}$ denotes an integer in $\mathcal{O}$ such that $\varepsilon_\mathfrak{p} \equiv -1 \bmod \mathfrak{p}^a$ and $\varepsilon_\mathfrak{p} \equiv +1 \bmod \mathfrak{a}\mathfrak{p}^{-a}$.*

If $\mathfrak{p} \nmid \mathfrak{m}$, then $\mathrm{E}(\underline{M}(\mathfrak{p}))$ is the trivial subgroup of $\mathrm{E}(\underline{M})$, otherwise $\mathrm{E}(\underline{M}(\mathfrak{p}))$ has order 2. (Recall by Proposition 1.13 that $\mathfrak{m}$ divides $\mathfrak{a}$.)

Proof The isomorphism follows from the Chinese remainder theorem (see for example [Neu99, I. 3, Theorem (3.6)]). It is obvious that the subgroup $\langle \varepsilon_\mathfrak{p} + \mathfrak{a} \rangle$ contains at most two elements.

If $\mathfrak{p} \nmid \mathfrak{m}$, then $\mathfrak{p}^a \| (2, \mathfrak{l})$, where $\mathfrak{l}$ is the level of $\underline{M}$. Here note by Proposition 1.13 that $\mathfrak{a} = \mathfrak{m}(2, \mathfrak{l})$. But this implies that $+1$ and -1 are equivalent modulo $\mathfrak{p}^a$, i.e. $\varepsilon_\mathfrak{p} \equiv +1 \bmod \mathfrak{a}$. Hence, $\mathrm{E}(\underline{M}(\mathfrak{p}))$ is the trivial subgroup of $\mathrm{E}(\underline{M})$.

If $\mathfrak{p}|\mathfrak{m}$, then $v_\mathfrak{p}(2, \mathfrak{l}) \leq a - 1$. But Proposition 1.13 implies that $v_\mathfrak{p}(2, \mathfrak{l}) = v_\mathfrak{p}(2\mathcal{O})$. Hence, $\mathfrak{p}^a \nmid 2$, i.e. $+1$ and -1 are inequivalent modulo $\mathfrak{p}^a$, and thus they are inequivalent modulo $\mathfrak{a}$, which implies finally that $\mathrm{E}(\underline{M}(\mathfrak{p}))$ has order 2. □

Proposition 1.23 *Let $\underline{M}$ be an $\mathcal{O}$-CM with level $\mathfrak{l}$, modified level $\mathfrak{m}$ and annihilator $\mathfrak{a}$. The linear characters of $\mathrm{E}(\underline{M})$ are parameterized by the square-free divisors of $\mathfrak{m}$. More precisely, the linear characters of $\mathrm{E}(\underline{M})$ are of the form:*

$$\psi_\mathfrak{f} : \mathrm{E}(\underline{M}) \to \{\pm 1\}, \quad \psi_\mathfrak{f}(\varepsilon + \mathfrak{a}) = \mu\big(\mathfrak{f}, (\varepsilon + 1)(2, \mathfrak{l})^{-1}\big),$$

where $\mathfrak{f}$ runs through the square-free divisors of $\mathfrak{m}$. (Here μ is the Möbius μ-function of K (see section Notations) and it is applied to the ideal $(\mathfrak{f}, \mathfrak{p})$.)

Remark Let $\mathfrak{p}^a \| \mathfrak{a}$ and $\mathfrak{p}|\mathfrak{m}$. If $\varepsilon = \varepsilon_\mathfrak{p}$, where $\varepsilon_\mathfrak{p}$ is as in Proposition 1.22, then we have $\psi_\mathfrak{f}(\varepsilon_\mathfrak{p} + \mathfrak{a}) = \mu(\mathfrak{f}, \mathfrak{p})$. Indeed, since $\mathfrak{a} = \mathfrak{m}(2, \mathfrak{l})$ (see Proposition 1.13) and $\mathfrak{p}|\mathfrak{m}$, we have $v_\mathfrak{p}(2, \mathfrak{l}) \leq a - 1$. However, $\varepsilon_\mathfrak{p} + 1$ is divisible by $\mathfrak{p}^a$. Hence, $\mathfrak{p}$ divides $(\varepsilon + 1)(2, \mathfrak{l})^{-1}$.

Proof of Proposition 1.23 To begin with, we show that $\psi_\mathfrak{f}$ is well-defined. First note that the ideal $(\varepsilon + 1)(2, \mathfrak{l})^{-1}$ is integral. Indeed, suppose that $\mathfrak{p}$ is an even prime ideal dividing $\mathfrak{l}$. Set $l = v_\mathfrak{p}(\mathfrak{l})$ and $u = v_\mathfrak{p}(2\mathcal{O})$. By Proposition 1.13, we have $u < l$, and hence $v_\mathfrak{p}(2, \mathfrak{l}) = u$. This implies that $\mathfrak{p}^u$ divides $(\varepsilon - 1)(\varepsilon + 1)$. Assume for contradiction that $\mathfrak{p}^u$ does not divide $\varepsilon + 1$. Then, say, $\mathfrak{p}^s$ divide $\varepsilon + 1$ $(s < u)$. Since $\mathfrak{p}^l$ divides $(\varepsilon - 1)(\varepsilon + 1)$, we have that $\varepsilon - 1$ is divisible by $\mathfrak{p}^{l-s}$. But $l - s > u$, since $l - s > l - u$ and $l - u \geq u$ (this is an easy consequence of Proposition 1.13). Hence, $\mathfrak{p}^u$ divides $\varepsilon - 1$. This is a contradiction, since $\varepsilon - 1 \equiv \varepsilon + 1 \bmod \mathfrak{p}^u$.

Now we show that the map $\psi_\mathfrak{f}$ depends only on the residue class of ε modulo $\mathfrak{a}$. Suppose $x \in \varepsilon + \mathfrak{a}$. We write $x = \varepsilon + a$, for some $a \in \mathfrak{a}$. Hence, we have $(x + 1)(2, \mathfrak{l})^{-1} = (\varepsilon + 1)(2, \mathfrak{l})^{-1} + a(2, \mathfrak{l})^{-1} \subseteq (\varepsilon + 1)(2, \mathfrak{l})^{-1} + \mathfrak{m}$ which proves well-definedness. Here we use the fact that $\mathfrak{a} = \mathfrak{m}(2, \mathfrak{l})$ (see Proposition 1.13).

Next we show that $\psi_\mathfrak{f}$ defines a group homomorphism from $\mathrm{E}(\underline{M})$ to $\{\pm 1\}$. Let $\varepsilon + \mathfrak{a}$, $\varepsilon' + \mathfrak{a} \in \mathrm{E}(\underline{M})$ and $\mathfrak{p}$ be a prime ideal divisor of $\mathfrak{m}$. We need to prove the following statements.

(i) If $\mathfrak{p}|(\varepsilon + 1)(2, \mathfrak{l})^{-1}$ and $\mathfrak{p}|(\varepsilon' + 1)(2, \mathfrak{l})^{-1}$, then $\mathfrak{p} \nmid (\varepsilon\varepsilon' + 1)(2, \mathfrak{l})^{-1}$
(ii) If $\mathfrak{p}|(\varepsilon + 1)(2, \mathfrak{l})^{-1}$ and $\mathfrak{p} \nmid (\varepsilon' + 1)(2, \mathfrak{l})^{-1}$, then $\mathfrak{p}|(\varepsilon\varepsilon' + 1)(2, \mathfrak{l})^{-1}$
(iii) If $\mathfrak{p} \nmid (\varepsilon + 1)(2, \mathfrak{l})^{-1}$ and $\mathfrak{p} \nmid (\varepsilon' + 1)(2, \mathfrak{l})^{-1}$, then $\mathfrak{p} \nmid (\varepsilon\varepsilon' + 1)(2, \mathfrak{l})^{-1}$.

We can write

$$\frac{\varepsilon\varepsilon' + 1}{(2, \mathfrak{l})} = \frac{(\varepsilon + 1)(\varepsilon' - 1)}{(2, \mathfrak{l})} + \frac{\varepsilon + 1}{(2, \mathfrak{l})} - \frac{\varepsilon' - 1}{(2, \mathfrak{l})} \tag{1.2}$$

$$= \frac{(\varepsilon - 1)(\varepsilon' + 1)}{(2, \mathfrak{l})} + \frac{\varepsilon' + 1}{(2, \mathfrak{l})} - \frac{\varepsilon - 1}{(2, \mathfrak{l})}. \tag{1.3}$$

First we prove that $\mathfrak{p}|(\varepsilon+1)(2,\mathfrak{l})^{-1}$ if and only if $\mathfrak{p} \nmid (\varepsilon-1)(2,\mathfrak{l})^{-1}$. If $\mathfrak{p}$ is odd, then this statement is obvious. If $\mathfrak{p}$ were even and the contrary held true, then $\mathfrak{p}$ would divide $2(2,\mathfrak{l})^{-1}$. But this is a contradiction, since $v_{\mathfrak{p}}(2\mathcal{O}) = v_{\mathfrak{p}}(2,\mathfrak{l})$ (see above). Now using this fact it is easy to deduce (i), (ii) (using (1.2)) and (iii) (using (1.3)).

It remains to show that every homomorphism χ from $\mathrm{E}(\underline{M})$ to $\{\pm 1\}$ is of this form, i.e. there exists a square-free divisor $\mathfrak{f}$ of $\mathfrak{m}$ such that $\chi(\varepsilon+\mathfrak{a}) = \psi_{\mathfrak{f}}(\varepsilon+\mathfrak{a})$ for any $\varepsilon+\mathfrak{a} \in \mathrm{E}(\underline{M})$. Let $\mathfrak{p}$ be a prime dividing $\mathfrak{m}$ and let $\varepsilon_{\mathfrak{p}}$ be as in Proposition 1.22. By setting

$$\mathfrak{f} = \prod_{\substack{\mathfrak{p}|\mathfrak{m} \\ \chi(\varepsilon_{\mathfrak{p}}+\mathfrak{a})=-1}} \mathfrak{p},$$

we recognize the claimed statement.

□

1.3 Some Lemmas Concerning Quotients $\mathcal{O}/\mathfrak{a}$

In this section $\mathfrak{a}$ stands for a nonzero integral ideal of $\mathcal{O}$. Moreover, R denotes the ring $\mathcal{O}/\mathfrak{a}$ and $\pi : \mathcal{O} \to R$ stands for the canonical projection.

In the present section we shall analyze the structure of the ring R and we shall provide some lemmas which will be needed in the next chapter.

Recall the well-known fact that the integral ideals of $\mathcal{O}$ containing $\mathfrak{a}$ are in one to one correspondence with the ideals of R via the map $\mathfrak{b} \mapsto \pi(\mathfrak{b})$.

Lemma 1.24 *If $\mathfrak{b}$ is an arbitrary integral ideal in $\mathcal{O}$, then one has $\pi(\mathfrak{b}) = \pi(\mathfrak{b}+\mathfrak{a})$.*

Proof It is clear that $\pi(\mathfrak{b}) \subseteq \pi(\mathfrak{b}+\mathfrak{a})$. Vice versa, let $\xi \in \pi(\mathfrak{b}+\mathfrak{a})$. Then $\xi = \pi(y+x)$ $(y \in \mathfrak{b}, x \in \mathfrak{a})$. Hence, $\xi = (y+x)+\mathfrak{a} = y+\mathfrak{a} = \pi(y) \subseteq \pi(\mathfrak{b})$. Therefore, the claimed identity holds true. □

Lemma 1.25 *The ring R is a principal ideal ring.*

Proof By the Chinese remainder theorem (see e.g. [Neu99, I. 3, Theorem (3.6)]), it suffices to consider $\mathfrak{a} = \mathfrak{p}^n$, where $\mathfrak{p}$ is a prime ideal of $\mathcal{O}$. We claim first of all that the ideals of $\mathcal{O}/\mathfrak{p}^n$ are $\pi(1), \pi(\mathfrak{p}), \ldots, \pi(\mathfrak{p}^n)$. Let I be an ideal of $\mathcal{O}/\mathfrak{p}^n$. Then $I = \pi(\pi^{-1}(I)) = \pi(\pi^{-1}(I)+\mathfrak{p}^n) = \pi(\mathfrak{p}^m)$ for some $0 \le m \le n$, i.e. the claim holds true. Note that the second equality follows from Lemma 1.24. For the third equality we used the fact that $\pi^{-1}(I)+\mathfrak{p}^n$ is an ideal of $\mathcal{O}$ containing $\mathfrak{p}^n$.

Next we prove that every ideal of $\mathcal{O}/\mathfrak{p}^n$ is principal. Fix an m such that $0 \le m \le n$. It suffices to prove that the ideal $\pi(\mathfrak{p}^m)$ is principal. Let $c \in \mathfrak{p}$ and $c \notin \mathfrak{p}^2$. Then $\mathfrak{p}^m = c^m\mathcal{O}+\mathfrak{p}^n$, since the greatest common divisor of c^m and $\mathfrak{p}^n$ is $\mathfrak{p}^m$. Hence, $\pi(\mathfrak{p}^m) = \pi(c^m\mathcal{O}+\mathfrak{p}^n) = \pi(c^m\mathcal{O}) = \pi(c^m)R$, where we used Lemma 1.24. This proves the lemma. □

Remark Let $\mathfrak{p}$ be a prime ideal of $\mathcal{O}$. The proof of Lemma 1.25 implies that the ideals of $\mathcal{O}/\mathfrak{p}^n$ are of the form $\pi(c^m)/\mathfrak{p}^n$ $(0 \leq m \leq n)$, where $c \in \mathfrak{p}$ and $c \notin \mathfrak{p}^2$, i.e. there are in total $n+1$ ideals of $\mathcal{O}/\mathfrak{p}^n$. From this we conclude that $\mathcal{O}/\mathfrak{p}^n$ is a principal local ring whose unique maximal ideal is $\pi(c)/\mathfrak{p}^n$.

Lemma 1.26 *We have* $\alpha R = \beta R$ *if and only if there exists some* $\varepsilon \in R^*$ *such that* $\alpha = \varepsilon\beta$.

Proof The statement would be trivial if R possessed no zero divisors, which, however does not hold true unless $\mathfrak{a}$ is prime. By the Chinese remainder theorem [Neu99, I. 3, Theorem (3.6)], it is enough to consider $\mathfrak{a} = \mathfrak{p}^n$, where $\mathfrak{p}$ is a prime ideal of $\mathcal{O}$. Let $\mathfrak{J}$ stand for the set of ideals of $\mathcal{O}/\mathfrak{p}^n$. We need to show that the following map

$$(\mathcal{O}/\mathfrak{p}^n)^*\backslash(\mathcal{O}/\mathfrak{p}^n) \to \mathfrak{J}, \quad [\alpha] \mapsto \alpha R$$

is an injection. For that it suffices to show that $(\mathcal{O}/\mathfrak{p}^n)^*\backslash(\mathcal{O}/\mathfrak{p}^n)$ has $n+1$ elements, since $\mathfrak{J}$ has $n+1$ elements (see Lemma 1.25 and the remark afterwards). (Here we use $[\alpha]$ for the orbit of α under left multiplication by elements of $(\mathcal{O}/\mathfrak{p}^n)^*$.)

Let $c \in \mathfrak{p}$ and $c \notin \mathfrak{p}^2$. It remains to prove the following identity

$$\sum_{m=0}^{n} |[c^m + \mathfrak{p}^n]| = \mathrm{N}(\mathfrak{p}^n), \tag{1.4}$$

since then we can take the elements $c^m + \mathfrak{p}^n$ $(0 \leq m \leq n)$ as representatives for the orbit space, i.e. the orbit space has $n+1$ elements as we claimed. We calculate the number of elements in each orbit, i.e. the number $|[c^m + \mathfrak{p}^n]|$ for all $0 \leq m \leq n$. By the so-called Orbit-Stabilizer theorem, we have $|[c^m + \mathfrak{p}^n]| = \varphi(\mathfrak{p}^n)/|\operatorname{Stab}(c^m + \mathfrak{p}^n)|$, where φ is the Euler's totient function, i.e. $\varphi(\mathfrak{a}) = |(\mathcal{O}/\mathfrak{a})^*|$ for an integral $\mathcal{O}$-ideal $\mathfrak{a}$. But we have

$$\begin{aligned}\operatorname{Stab}(c^m + \mathfrak{p}^n) &= \{x + \mathfrak{p}^n \in (\mathcal{O}/\mathfrak{p}^n)^* : xc^m \equiv c^m \bmod \mathfrak{p}^n\} \\ &= \{x + \mathfrak{p}^n \in (\mathcal{O}/\mathfrak{p}^n)^* : x \equiv 1 \bmod \mathfrak{p}^{n-m}\}.\end{aligned}$$

The above identity implies that $\operatorname{Stab}(c^m + \mathfrak{p}^n)$ equals the kernel of the reduction map $(\mathcal{O}/\mathfrak{p}^n)^* \to (\mathcal{O}/\mathfrak{p}^{n-m})^*$. From the first isomorphism theorem for groups, we have $|\operatorname{Stab}(c^m + \mathfrak{p}^n)| = \varphi(\mathfrak{p}^n)/\varphi(\mathfrak{p}^{n-m})$. Therefore, $|[c^m + \mathfrak{p}^n]| = \varphi(\mathfrak{p}^{n-m})$. Hence, to obtain (1.4), it is enough to show $\varphi(\mathfrak{p}^n) = \mathrm{N}(\mathfrak{p}^n) - \mathrm{N}(\mathfrak{p}^{n-1})$. But from Lemma 1.25 and the remark afterwards we have that $\mathcal{O}/\mathfrak{p}^n$ is a local principal ideal ring with the maximal ideal $\pi(c)/\mathfrak{p}^n$, where $c \in \mathfrak{p}$ and $c \notin \mathfrak{p}^2$. Since $\pi(c)/\mathfrak{p}^n$ has $\mathrm{N}(\mathfrak{p}^{n-1})$-many elements, i.e. the non-units in $\mathcal{O}/\mathfrak{p}^n$ are $\mathrm{N}(\mathfrak{p}^{n-1})$ in total, the last assertion holds true. □

Remark Let $\mathfrak{p}$ be a prime ideal of $\mathcal{O}$. From the proof of Lemma 1.26 we have that the elements $c^m + \mathfrak{p}^n$ $(0 \leq m \leq n)$, where c is an element in $\mathfrak{p}$ but not in $\mathfrak{p}^2$, can be taken as representatives for the orbit space of the left action of the group $(\mathcal{O}/\mathfrak{p}^n)^*$ on $(\mathcal{O}/\mathfrak{p}^n)$.

Lemma 1.27 *If $\mathcal{O} = x\mathcal{O} + y\mathcal{O} + \mathfrak{a}$, then there exists $x', y' \in \mathcal{O}$ such that $x' \equiv x \bmod \mathfrak{a}$ and $y' \equiv y \bmod \mathfrak{a}$ with $\mathcal{O} = x'\mathcal{O} + y'\mathcal{O}$.*

Proof The statement is obvious if $x = y = 0$. Without loss of generality we assume $y \neq 0$. Let $\mathfrak{g} := x\mathcal{O} + y\mathcal{O}$. We set

$$\mathfrak{y}_1 := \prod_{\substack{\mathfrak{p}^t \| y\mathcal{O} \\ \mathfrak{p} | \mathfrak{g}}} \mathfrak{p}^t, \qquad \mathfrak{y}_2 := (y\mathcal{O})\mathfrak{y}_1^{-1}.$$

Obviously, $\mathfrak{y}_1$ and $\mathfrak{y}_2$ are relatively prime. Let $\mathfrak{t}$ be an integral $\mathcal{O}$-ideal in the inverse ideal class of $\mathfrak{y}_2\mathfrak{a}$ which is relatively prime to $xy\mathfrak{a}$. Then there exists $a \in \mathcal{O}$ such that $a\mathcal{O} = \mathfrak{y}_2\mathfrak{a}\mathfrak{t}$. Set $x' = x + a$, $y' = y$. It remains to show that $\mathfrak{h} := x'\mathcal{O} + y\mathcal{O}$ equals $\mathcal{O}$.

Assume for contradiction that there exists a prime ideal $\mathfrak{p}$ dividing $\mathfrak{h}$. Then $\mathfrak{p}$ divides either $\mathfrak{y}_1$ or $\mathfrak{y}_2$. If $\mathfrak{p}$ divides $\mathfrak{y}_1$, then by the very definition of $\mathfrak{y}_1$, the ideal $\mathfrak{p}$ divides $\mathfrak{g}$, and hence it divides x. But since $\mathfrak{p}$ also divides x', the ideal $\mathfrak{p}$ divides $a\mathcal{O} = \mathfrak{y}_2\mathfrak{a}\mathfrak{t}$. But by the choice of $\mathfrak{t}$, the ideals $\mathfrak{p}$ and $\mathfrak{t}$ are relatively prime, hence $\mathfrak{p}$ divides $\mathfrak{a}$ which contradicts with the assumption.

If $\mathfrak{p}$ divides $\mathfrak{y}_2$, then $\mathfrak{p}$ divides $a\mathcal{O}$, and hence it divides $x\mathcal{O}$. But this implies that $\mathfrak{p}$ divides $\mathfrak{g}$, and hence it divides $\mathfrak{y}_1$ which contradicts with the fact that $\mathfrak{y}_1$ and $\mathfrak{y}_2$ are relatively prime. This proves the lemma. □

Lemma 1.28 *If $R = \alpha R + \beta R$, then there exists $x \in \alpha$ and $y \in \beta$ such that $\mathcal{O} = x\mathcal{O} + y\mathcal{O}$.*

Proof Since π is a surjection, we have $\pi(\mathcal{O}) = \pi(x)\pi(\mathcal{O}) + \pi(y)\pi(\mathcal{O})$, where $\alpha = \pi(x)$ and $\beta = \pi(y)$ with $x, y \in \mathcal{O}$. Then we have $\pi(\mathcal{O}) = \pi(x\mathcal{O} + y\mathcal{O})$, i.e. $\mathcal{O} = x\mathcal{O} + y\mathcal{O} + \mathfrak{a}$. Lemma 1.27 implies now the result. □

Lemma 1.29 *The map $\underline{\pi} : \mathrm{SL}(2, \mathcal{O}) \to \mathrm{SL}(2, R)$, $\left(\begin{smallmatrix} a & b \\ c & d \end{smallmatrix}\right) \mapsto \left(\begin{smallmatrix} \pi(a) & \pi(b) \\ \pi(c) & \pi(d) \end{smallmatrix}\right)$ defines an epimorphism.*

Proof We write Γ and Γ_R for $\mathrm{SL}(2, \mathcal{O})$ and $\mathrm{SL}(2, R)$, respectively. It is clear that the map $\underline{\pi}$ is a group homomorphism. Let $B := \left(\begin{smallmatrix} \pi(a) & \pi(b) \\ \pi(c) & \pi(d) \end{smallmatrix}\right) \in \Gamma_R$ with $a, b, c, d \in \mathcal{O}$. Here note that since π is a surjection, every element of Γ_R is of this form. To prove the lemma we need to find an element A in Γ such that $\underline{\pi}(A) = B$.

Since $B \in \Gamma_R$, we have $\pi(c)R + \pi(d)R = R$. From Lemma 1.28, there exists x in $\pi(c)$ and y in $\pi(d)$ such that $x\mathcal{O} + y\mathcal{O} = \mathcal{O}$. Since $\pi(x) = \pi(c)$ and $\pi(y) = \pi(d)$, we have $\pi(ad - bc) = \pi(ay - bx)$. Hence, we can write $ay - bx = 1 + k(x - y)$ for some $k \in \mathfrak{a}$. Hence, $(a + k)y - (b + k)x = 1$. Therefore, the matrix $\left(\begin{smallmatrix} a+k & b+k \\ x & y \end{smallmatrix}\right)$ is an element of Γ and, obviously, it is mapped to B by $\underline{\pi}$. This proves the lemma. □

Lemma 1.30 *Given $\alpha, \beta \in R$. There exists $\gamma \in R$ and $A \in \mathrm{SL}(2, R)$ such that $(0, \gamma)A = (\alpha, \beta)$. Here multiplication of a row vector with a matrix is done in the usual way.*

Proof By the Chinese remainder theorem [Neu99, I. 3, Theorem (3.6)] it is enough to consider $\mathfrak{a} = \mathfrak{p}^n$. Let $a \in \alpha$ and $b \in \beta$. From the remark after Lemma 1.26, we can write $a \equiv c^{m_1} e_1 \mod \mathfrak{p}^n$ and $b \equiv c^{m_2} e_2 \mod \mathfrak{p}^n$ $(0 \leq m_1, m_2 \leq n)$, where $e_1, e_1 \in (\mathcal{O}/\mathfrak{p}^n)^*$.

If $m_1 \leq m_2$, we have

$$(0, c^{m_1}) \begin{pmatrix} 0 & -e_1^{-1} \\ e_1 & c^{m_2 - m_1} e_2 \end{pmatrix} = (c^{m_1} e_1, c^{m_2} e_2) \equiv (a, b) \bmod \mathfrak{p}^n.$$

By taking $\gamma = \pi(c^m)$ and $A = \begin{pmatrix} 0 & -\pi(e_1)^{-1} \\ \pi(e_1) & \pi(c^{m_2 - m_1} e_2) \end{pmatrix}$ we see that the statement of the lemma holds true.

If $m_2 \leq m_1$, then using the above argument, we find $A \in \mathrm{SL}(2, \mathcal{O}/\mathfrak{p}^n)$ and γ in $\mathcal{O}/\mathfrak{p}^n$ such that $(0, \gamma)A = (\beta, -\alpha)$. Since $(\beta, -\alpha) = (\alpha, \beta)S$, where $S = \begin{pmatrix} 0 & -1 \\ 1 & 0 \end{pmatrix}$, we obtain $(0, \gamma)AS^{-1} = (\alpha, \beta)$. This proves the lemma. □

Chapter 2
Weil Representations of Finite Quadratic Modules

We carry over the notations of the previous chapter. As before, K denotes a number field of degree n over $\mathbb{Q}$, and we use $\mathcal{O}$, $\mathfrak{d}$ for the ring of integers and the different of K, respectively. Moreover, we shall use Γ for the group $\mathrm{SL}(2, \mathcal{O})$ and $\tilde{\Gamma}$ for a certain central extension of Γ (see Sect. 2.2 for the definition of $\tilde{\Gamma}$). Occasionally, we shall denote by Γ_R for a ring R, the group $\mathrm{SL}(2, R)$.

In this chapter we shall associate Weil representations to finite quadratic $\mathcal{O}$-modules ($\mathcal{O}$-FQM) and develop a basic theory of these representations. The main result of this chapter will be Theorem 2.5, which describes the complete decomposition of cyclic Weil representations, i.e of Weil representations associated to *cyclic* $\mathcal{O}$-FQM, and Theorem 2.6, which gives the explicit description of all one-dimensional subrepresentations of cyclic Weil representations. The latter theorem will play an important role when we determine explicitly all singular Jacobi forms of certain indices, since in Chap. 4, we shall see that the singular Jacobi forms will correspond to the one-dimensional subrepresentations of certain cyclic Weil representations.

In Sect. 2.1, we shall briefly recall notations and facts concerning representations of groups which will be used in the sequel. In Sect. 2.2, we shall define the Weil representations of $\tilde{\Gamma}$ associated to $\mathcal{O}$-FQM. In fact, though we shall use throughout the term *representations* we view the Weil representations rather as modules over $\tilde{\Gamma}$. In Sect. 2.3, we shall study decompositions of Weil representations. For arbitrary finite quadratic modules, our decompositions of the associated Weil representations are in general not complete, i.e. they are neither splittings into direct sums nor the subrepresentations which occur in the given decompositions are irreducible. However, as we shall see in Sect. 2.4, for *cyclic* Weil representations, these decompositions are in fact complete. The proof of this completeness relies on Theorem 2.7 of Sect. 2.6, which provides an upper bound for the number of irreducible $\tilde{\Gamma}$-submodules of an arbitrary Weil representation. Using the dimension formulas for the irreducible $\tilde{\Gamma}$-submodules of a cyclic Weil representation, we shall

H. Boylan, *Jacobi Forms, Finite Quadratic Modules and Weil Representations over Number Fields*, Lecture Notes in Mathematics 2130,
DOI 10.1007/978-3-319-12916-7_2

be able to determine in Sect. 2.5, the one-dimensional submodules of cyclic Weil representations.

In Sect. 2.6, for deriving our estimate for the number of irreducible $\tilde{\Gamma}$-submodules of cyclic Weil representations, we have to develop a machinery which is also interesting for its own sake since it gives a explicit description of the Weil representations as projective representations of $\Gamma = \mathrm{SL}(2, \mathcal{O})$.

2.1 Review of Representations of Groups

To fix the language, we shall briefly recall those basic notions and facts from the general theory of representations of groups which we shall need in the sequel. In particular, we shall introduce and discuss the notion of a projective action of a group on a vector space. This notion will be useful for us in later sections.

In the following, G denotes a multiplicative group with identity element 1. For the convenience of the reader we shall give proofs of some basic propositions for which we could not find suitable references.

Definition 2.1 Let G be a group acting from left on a set X. We use $G\backslash X$ for the set of orbits of the G-action. For an element $v \in X$, we use $\mathrm{Stab}(v)$ for the subset of elements of G which are fixed under the G-action, i.e. $\mathrm{Stab}(v) = \{g \in G : gv{=}v\}$. (In fact, the set $\mathrm{Stab}(v)$ is a subgroup of G.) If G acts from right on X, the set of orbits of the G-action is denoted by X/G.

Unless otherwise stated when we speak of a group action, we suppose that the group acts from the left.

Proposition 2.2 *Suppose there exits a group homomorphism π from G onto H. If H acts on a set X, then G also acts on X, via $gv := \pi(g)v$. If π is surjective then the number of elements of $G\backslash X$ equals the number of elements of $H\backslash X$.*

Proof The proof is obvious. □

Definition 2.3 Let f be a group homomorphism from G to G' and π be a surjective group homomorphism from G onto H. We say that *f factors through π* (or sometimes, if π is clear from the context, that *f factors through H*), if there exits a group homomorphism $\underline{f}$ from H to G' such that $\underline{f} \circ \pi = f$.

The following proposition is a standard proposition from basic algebra.

Proposition 2.4 *Let G, G', H, π and f be as in Definition 2.3. Then f factors through π if and only if* $\mathrm{Ker}(\pi) \subseteq \mathrm{Ker}(f)$.

Proof If $\underline{f}$ exists such that $\underline{f} \circ \pi = f$ then obviously $\mathrm{Ker}(\pi) \subseteq \mathrm{Ker}(f)$. Assume vice versa $\mathrm{Ker}(\pi) \subseteq \mathrm{Ker}(f)$. Let $h \in H$ be given. Since π is a surjection, there exists $g \in G$ such that $\pi(g) = h$. We define $\underline{f} : H \to G'$ by $\underline{f}(h) = f(g)$. By the assumption this map is well-defined, i.e. does not depend on the particular choice of the preimage g of h. By definition $\underline{f}(\pi(g)) = f(g)$. Finally it is obvious that $\underline{f}$ is a group homomorphism. □

Definition 2.5 A *G-module* is a vector space V together with an operation from $G \times V$ to V of G on V such that, for each g in G, the map $v \mapsto gv$ is linear. If the action is clear from the context then we simply say that *V is a G-module*. The group homomorphism

$$\rho : G \to \mathrm{GL}(V), \quad \rho(g)(v) = gv$$

is called the *representation afforded by V*. If G acts from the right on V, then we say that *V is a right G-module*. (In this case ρ satisfies $\rho(gh) = \rho(h)\rho(g)$ for all g, h in G.)

Remark If there exists a representation of G on V, i.e. a group homomorphism $\rho : G \to \mathrm{GL}(V)$, then V becomes a G-module via the action $(g, v) \mapsto gv := \rho(g)(v)$ and ρ is the representation afforded by this V-module.

Definition 2.6 Let V be a G-module and ρ be the representation afforded by V. Suppose that there exists a surjective group homomorphism π from G onto H. If ρ factors through π, then we say that the *representation of G factors through a representation of H* and that the *G-action on V factors through an action of H*.

Definition 2.7 Let V be a finite dimensional complex Hilbert space with inner product $\langle\ ,\ \rangle$. We call a representation $\rho : G \to \mathrm{GL}(V)$ unitary if $\rho(g)$ is unitary for every g in G (i.e. if $\langle \rho(g)v, \rho(g)w\rangle = \langle v, w\rangle$ for all v, w in V). We say that *G acts unitarily on V*, if the afforded representation is unitary.

Definition 2.8 Let V be G-module and ρ be the representation afforded by V. A subspace W of V is called a *G-invariant subspace of V* or a *G-submodule of V*, if $gW \subseteq W$ $(g \in G)$. The G-module V or the representation ρ is called *irreducible*, if there is no proper nonzero G-invariant subspace of V.

Definition 2.9 Let V be a G-module and ρ be the *representation afforded by V*. For $g \in G$, we use $\mathrm{tr}(g, V)$ for the trace of the matrix determined by the automorphism $\rho(g)$ of V. The map $\chi_V : G \to \mathbb{C}$ defined by $\chi_V(g) = \mathrm{tr}(g, V)$ is called the *character of the G-module V* and the *character or trace of ρ*. An *irreducible character of G* is a character of an irreducible G-module. A *linear character of G* is a character of a one-dimensional G-module.

Remark Every linear character of a finite group G gives rise to a group homomorphism from G to $\mathbb{S}^1$, and vice versa.

Proposition 2.10 *If $V = \bigoplus_{i=1}^n V_i$ is a decomposition of V into G-invariant subspaces, then the character of V equals the sum of the characters of V_i, i.e. we have:*

$$\chi_V(g) = \sum_{i=1}^n \chi_{V_i}(g) \quad (g \in G).$$

Proof We proceed by induction on n. For $n = 1$, the result is immediate. We prove the result for $n = 2$. Write $V = A \bigoplus B$, where A and B are G-invariant subspaces of V. Let $\{a_1, \ldots, a_s\}$ be a basis for A and $\{b_1, \ldots, b_t\}$ be a basis for B, and let g in G. Since A and B are G-submodules of V, we have $ga_i \in A$ and $gb_j \in B$, and therefore $ga_i = \sum_{k=1}^{s} c_{ki} a_j$ and $gb_j = \sum_{l=1}^{t} d_{lj} b_l$ with $c_{ki}, d_{lj} \in \mathbb{C}$. But then $\operatorname{tr}(g, A) = \sum_k c_{kk}$, $\operatorname{tr}(g, B) = \sum_l d_{ll}$ and $\operatorname{tr}(g, V) = \sum_k c_{kk} + \sum_l d_{ll}$, which is the claimed formula. Suppose $n \geq 2$. Using the previous result, we have $\operatorname{tr}(g, V) = \operatorname{tr}(g, V_n) + \operatorname{tr}\left(g, \bigoplus_{i=1}^{n-1} V_i\right)$. By induction hypothesis, we have $\operatorname{tr}\left(g, \bigoplus_{i=1}^{n-1} V_i\right) = \bigoplus_{i=1}^{n-1} \operatorname{tr}(g, V_i)$. This proves the proposition. □

Definition 2.11 Let V and W be G-modules. A $\mathbb{C}$-linear map $\varphi : V \to W$ is called *G-linear*, if $\varphi(gv) = g\varphi(v)$ $(g \in G, v \in V)$.

Definition 2.12 Let V and W be G-modules. Let ρ and σ stand for the representations afforded by V and W, respectively. We say that V and W are isomorphic as G-modules, and that ρ and σ are *equivalent*, if there exists a G-linear isomorphism $\tau : V \to W$ (or, equivalently, if there exists an isomorphism of vector spaces $\tau : V \to W$ such that $\tau \circ \rho(g) = \sigma(g) \circ \tau$ for all $g \in G$).

Proposition 2.13 ([FH91, Cor. 2.13, Cor. 2.14]) *Let G be a finite group. The set of irreducible characters of G is finite. Two G-modules are isomorphic as G-modules if their characters coincide.*

Definition 2.14 For a G-module V we use V^G for the *subspace of G-invariants of V*, i.e. the subspace of all v in V such tat $gv = v$ for all g in G.

Proposition 2.15 *Let G be a finite group. If V is a G-module, then the application $v \mapsto \frac{1}{|G|} \sum_{g \in G} gv$ defines a surjective map $\varphi : V \to V^G$.*

Proof It is clear that the map is well defined, i.e. $\varphi(V) \subseteq V^G$. Suppose $v \in V^G$. Then we have $gv = v$ for all g. Hence $\varphi(v) = v$, which implies that the map is surjective. □

Proposition 2.16 *Let V be a G-module. Suppose $V = \bigoplus_{i=1}^{r} V_i$ with irreducible G-submodules V_i. If W is a nonzero irreducible G-submodule of V, then W is G-linearly isomorphic to V_i for some i.*

Proof Let P_i denote the projection of V onto V_i. Then $\sum_{i=1}^{r} P_i = 1$. Since $W \neq 0$, there exists some i such that $P_i(W) \neq 0$. Since $P_i(W)$ is a G-submodule of V_i, and V_i is irreducible, we have $P_i(W) = V_i$. So, the map $P_i|_W$ is surjective. But the kernel of $P_i|_W$ is a G-submodule of W, so it must be zero, since W is irreducible and $P_i|_W \neq 0$. This proves the proposition. □

Proposition 2.17 ([FH91, Prop. 1.8]) *Let G be finite, let $\hat{G}$ denote the set of irreducible characters of G. For χ in $\hat{G}$, let V^χ denote the sum of those G-submodules of V which afford the character χ. Then V^χ is the largest G-submodule of V whose character is a multiple of χ. Moreover, one has*

$$V = \bigoplus_{\chi \in \hat{G}} V^{\chi}. \tag{2.1}$$

If $V = \bigoplus_{j=1}^{m} W_j$ is a decomposition of V into G-submodules W_j such that the character of W_j is a multiple of an irreducible character χ_j and such that $\chi_j \neq \chi_k$ for $j \neq k$, then $W_j = V^{\chi_j}$, and the splitting $V = \bigoplus W_j$ coincides with the decomposition (2.1) *(after deleting zero spaces up to permutation of the summands).*

Definition 2.18 The decomposition (2.1) is called the *canonical decomposition of the G-module V.*

Remark If G is abelian, then $\hat{G}$ is a group with respect to the usual multiplication of functions. Namely, $\hat{G}$ coincides with the group of linear characters of G, called the *dual group of G*. In this case, for $\chi \in \hat{G}$, we have $V^{\chi} = \sum_v \mathbb{C}v$, where the sum is over all $v \in V$ which satisfy $gv = \chi(g)v$ for all $g \in G$. In other words, we have

$$V^{\chi} = \{v \in V : gv = \chi(g)v, \forall g \in G\}.$$

Proposition 2.19 *We carry over the notations of Proposition 2.17. Let V_i be an irreducible G-submodule of V^{χ}. Then,*

$$\dim V^{\chi} = \frac{\dim V_i}{|G|} \sum_{g \in G} \overline{\chi_{V_i}(g)} \chi_V(g).$$

Proof From [FH91, Eq. (2.32)], we know that $\pi : \dim V_i/|G| \sum_{g\in G} \overline{\chi_{V_i}(g)} g$ defines a projection from V onto V^{χ}. Let $v_1, \ldots, v_m$ be a basis for V^{χ} and $v_{m+1}, \ldots, v_n$ be a basis for the kernel of π. Hence, $v_1, \ldots, v_n$ becomes a basis for V. We then have $\pi(v_i) = v_i$ for $1 \leq i \leq m$ and $\pi(v_i) = 0$ otherwise. Hence $\dim V^{\chi}$ equals the trace of the map π. This proves the proposition. □

Corollary 2.20 *Let G be a finite group and V be a G-module. We have*

$$\dim V^{G} = \frac{1}{|G|} \sum_{g \in G} \chi_V(g).$$

Proof This follows immediately from Proposition 2.19 in the case of χ being the trivial character. □

Let V and W be G-modules. The spaces $V \oplus W$ and $V \otimes W$ are also G-modules via

$$(g, v \oplus w) \mapsto (gv \oplus gw), \quad (g, v \otimes w) \mapsto (gv \otimes gw), \tag{2.2}$$

respectively. The space of all $\mathbb{C}$-linear maps from V to W is denoted by $\mathrm{Hom}(V, W)$. In particular, we set $V^* := \mathrm{Hom}(V, \mathbb{C})$. Moreover, the space of all G-linear maps from V to W is denoted by $\mathrm{Hom}_G(V, W)$. In particular, the space $\mathrm{Hom}_G(V, V)$ is called the *intertwining algebra of* V. It is not difficult to see that $\mathrm{Hom}(V, W)$ is a G-module via the following G-action

$$(g, \lambda) \mapsto {}^g\lambda, \quad {}^g\lambda(v) := g\lambda(g^{-1}v). \tag{2.3}$$

In fact, (2.3) defines also a G-module structure on $\mathrm{Hom}(V, W)$ if $V = W$ and if V is not a G-module but only a projective G-module (see Definition 2.23 below), provided the projective representation ρ afforded by V satisfies $\rho(g^{-1})\rho(g) = 1$.

We denote by $V^\bullet$ the G-module whose underlying vector space is the dual space V^* of V and where the G-action is given by:

$$(g, \lambda) \mapsto {}^g\lambda \quad \text{where } {}^g\lambda(v) = \lambda(g^{-1}v). \tag{2.4}$$

The spaces $V^\bullet \otimes W$ and $\mathrm{Hom}(V, W)$ can be identified via the following G-module isomorphism

$$V^\bullet \otimes W \to \mathrm{Hom}(V, W), \quad \lambda \otimes \omega \mapsto \text{“}v \mapsto \lambda(v)\omega\text{''}. \tag{2.5}$$

Proposition 2.21 *Let V, W be G-modules. The following holds true:*

$$\mathrm{Hom}_G(V, W) = \mathrm{Hom}\left(V, W\right)^G.$$

Proof From (2.3), for any $g \in G$ and any $v \in V$, we have ${}^g\lambda(v) = g\lambda(g^{-1}v)$. So, if λ is G-linear (i.e. $\lambda \in \mathrm{Hom}_G(V, W)$), then clearly ${}^g\lambda = \lambda$ (i.e. $\lambda \in \mathrm{Hom}\left(V, W\right)^G$). On the other hand, if ${}^g\lambda = \lambda$, then clearly $g^{-1}\lambda(v) = \lambda(g^{-1}v)$ which implies that λ is G-linear. This proves the proposition. □

Proposition 2.22 *Let G be a finite group, and V be a G-module. The number of irreducible G-submodules of V is less than or equal to the dimension of the space $\mathrm{Hom}_G(V, V)$.*

Proof We denote by χ_j $(j = 1 \ldots, m)$ the characters of the distinct irreducible G-submodules of V (see Proposition 2.13 for finiteness of the number of irreducible characters). Let $V = \oplus_{j=1}^m V^{\chi_j}$ be the canonical decomposition of V (see Proposition 2.17), where V^{χ_j} is the sum of those G-submodules of V which have character χ_j. It is enough to prove

$$\mathrm{Hom}_G(V, V) \simeq \oplus_{j=1}^m \mathrm{Hom}_G(V^{\chi_j}, V^{\chi_j}). \tag{2.6}$$

Namely, let $V^{\chi_j} = \bigoplus_{k=1}^{d_j} V_{j,k}$ be a decomposition into irreducible submodules. Then, from the fact $\dim \mathrm{Hom}_G(V^{\chi_j}, V^{\chi_j}) = d_j^2$ which we shall prove in a moment, we obtain $\dim \mathrm{Hom}_G(V, V) \geq m$ (since $d_j^2 \geq 1$ for all j).

First we prove $\dim \mathrm{Hom}_G(V^{\chi_j}, V^{\chi_j}) = d_j^2$. From Proposition 2.21, we have $\mathrm{Hom}_G(V^{\chi_j}, V^{\chi_j}) = \mathrm{Hom}(V^{\chi_j}, V^{\chi_j})^G$. Using Corollary 2.20, we then have

$$\dim \mathrm{Hom}_G(V^{\chi_j}, V^{\chi_j}) = \frac{1}{|G|} \sum_{g \in G} \chi_{\mathrm{Hom}(V^{\chi_j}, V^{\chi_j})}(g).$$

But via the identification in (2.5) and [FH91, Prop. 2.1], we have

$$\chi_{\mathrm{Hom}(V^{\chi_j}, V^{\chi_j})} = \overline{\chi_{V^{\chi_j}}}\chi_{V^{\chi_j}} = d_j^2 \overline{\chi}_j \chi_j.$$

Using [FH91, Eq. (2.10)], we now recognize the claimed identity.

Finally, we prove (2.6). Let $L \in \mathrm{Hom}_G(V, V)$. If we can show that for each j, $L(V^{\chi_j})$ is a subset of V^{χ_j}, then obviously the map $L \mapsto (L|_{V^{\chi_j}})_j$ defines an isomorphism. Since L is a linear map, we can write $L(V^{\chi_j}) = \sum_{k=1}^{d_j} L(V_{j,k})$. The kernel of $L|_{V_{j,k}}$ is either 0 or $V_{j,k}$, since $V_{j,k}$ is an irreducible G-module. If the kernel of $L|_{V_{j,k}}$ is $V_{j,k}$, there is nothing to prove. Suppose the kernel of $L|_{V_{j,k}}$ is 0. Then $L(V_{j,k}) \simeq V_{j,k}$, and hence $L(V_{j,k}) \subseteq V^{\chi_j}$. But this implies that $L(V^{\chi_j}) \subseteq V^{\chi_j}$. □

Definition 2.23 A *projective action of G on a vector space V* is a map $G \times V \to V$, $(g, v) \mapsto gv$ such that, for all $g, h \in G$, there exists a constant $\lambda(g, h) \in \mathbb{C}^*$ such that

(i) $g(hv) = \lambda(g, h)(gh)v \quad (v \in V, g, h \in G)$,
(ii) $1v = v \quad (v \in V)$.

The space V is then called a *projective G-module*. The map $\rho : G \to \mathrm{GL}(V)$, $\rho(g) = gv$ is called the *projective representation afforded by the projective G-module V*. The map $\lambda : G \times G \to \mathbb{C}^*$ is called the *multiplier system of the projective G-module V*.

Remark Note that the projective representation afforded by the G-module V satisfies $\rho(g)\rho(h) = \lambda(g, h)\rho(gh)$ for all g, h in G. If vice versa ρ is a *projective representation of G*, i.e. if ρ is a map $\rho : G \to \mathrm{GL}(V)$ such that for all $g, h \in G$ there exists a constant $\lambda(g, h) \in \mathbb{C}^*$ which satisfies

$$\rho(g)\rho(h) = \lambda(g, h)\rho(gh) \quad (g, h \in G),$$

then the map $(g, v) \mapsto \rho(g)(v)$ defines a projective G-module structure on V and ρ is the projective representation afforded by V.

Remark Note that the multiplier system of a projective G-module satisfies

$$\lambda(1, g) = \lambda(g, 1) = 1 \tag{2.7}$$

$$\lambda(g', g'')\lambda(g, g'g'') = \lambda(g, g')\lambda(gg', g''), \tag{2.8}$$

as follows immediately from the axioms (i) and (ii). Indeed, for proving (2.8), we write, for $v \in V$

$$g\big(g'(g''v)\big) = \lambda(g', g'')g\big((g'g'')v\big) = \lambda(g', g'')\lambda(g, g'g'')(gg'g'')v$$
$$g\big(g'(g''v)\big) = \lambda(g, g')(gg')(g''v) = \lambda(g, g')\lambda(gg', g'')(gg'g'')v,$$

and comparing yields the claimed cocycle relation.

Definition 2.24 Let V be a projective G-module with multiplier system λ. We define G_V to be the set

$$G_V = \{(g, \xi) : g \in G, \xi \in C\},$$

where C is the subgroup of $\mathbb{C}^*$ generated by the $\lambda(g, h)$ $(g, h \in G)$ together with the multiplication

$$(g, \xi) \cdot (g', \xi') := \big(gg', \xi\xi'\lambda(g, g')\big). \tag{2.9}$$

Proposition 2.25 *The multiplication* (2.9) *defines the structure of a group on* G_V. *The sequence*

$$1 \to C \xrightarrow{\iota} G_V \xrightarrow{\pi} G \to 1,$$

where $\iota(\xi) = (1, \xi)$ *and* $\pi(g, \xi) = g$, *and the subgroup* $\iota(C)$ *lies in the center of* G_V. *In short,* G_V *is a central extension of* G *by the abelian group* C.

Proof First we show that G_V becomes a group with the operation given in (2.9). The neutral element of G_V is $(1, 1)$ as follows from (2.7). The inverse of an arbitrary element (g, ξ) is $\big(g^{-1}, \xi^{-1}\lambda(g, g^{-1})^{-1}\big)$.

It remains to prove the associativity of the multiplication, i.e. we need to prove

$$(g, \xi) \cdot \big((g', \xi') \cdot (g'', \xi'')\big) = \big((g, \xi) \cdot (g', \xi')\big) \cdot (g'', \xi'').$$

On the left we have

$$\big(gg'g'', \xi\xi'\xi''\lambda(g', g'')\lambda(g, g'g'')\big).$$

On the right we have

$$\big(gg'g'', \xi\xi'\xi''\lambda(g, g')\lambda(gg', g'')\big).$$

Applying (2.8) shows that both sides coincide.

The exactness of the given sequence is obvious. That $\iota(C)$ lies in the center of G_V follows again from (2.7). □

Proposition 2.26 *Let V be a projective G-module. The space V becomes a G_V-module via the following G_V-action:*

$$\big((g,\xi),v\big) \mapsto (g,\xi)v := \xi(gv).$$

Proof Clearly $(1,1)$ acts as identity. Let $(g,\xi),(g',\xi') \in G_V$. For checking the second axiom for a G-action we calculate

$$\begin{aligned}\big((g,\xi)\cdot(g',\xi')\big)v &= \big(gg',\xi\xi'\lambda(g,g')\big)v = \xi\xi'\lambda(g,g')gg'v\\ &= \xi\xi'\big(g(g'v)\big) = \xi\big(g(\xi'g'v)\big) = (g,\xi)\big((g',\xi')v\big).\end{aligned}$$

This proves the proposition. □

2.2 The Weil Representation $W(\underline{M})$

Theorem 2.1 *The group $\Gamma = \mathrm{SL}(2,\mathcal{O})$ is generated by $S = \left(\begin{smallmatrix}0 & -1\\ 1 & 0\end{smallmatrix}\right)$ and $T_b = \left(\begin{smallmatrix}1 & b\\ 0 & 1\end{smallmatrix}\right)$ $(b \in \mathcal{O})$.*

Proof A theorem of Vaserstein [Vas72, First Thm.] states that Γ is generated by the matrices $T^c := \left(\begin{smallmatrix}1 & 0\\ c & 1\end{smallmatrix}\right)$ and T_b $(b,c \in \mathcal{O})$. However, we have the following easily deduced identity:

$$T^{-b} = ST_bS^{-1}.$$

□

Let $\underline{M} = (M,Q)$ be an $\mathcal{O}$-FQM with associated bilinear form B. We use $\mathbb{C}[M]$ for the complex vector space of maps $M \to \mathbb{C}$. Recall from the section Notations that the functions e_x $(x \in M)$ form a basis of $\mathbb{C}[M]$. To the generators of Γ from Theorem 2.1 we assign linear operators $U(S)$ and $U(T_b)$ on $\mathbb{C}[M]$ by setting for the basis elements e_x:

$$\begin{aligned}U(T_b)e_x &= e\{bQ(x)\}\,e_x \quad (b \in \mathcal{O})\\ U(S)e_x &= \sigma(\underline{M})\frac{1}{\sqrt{|M|}}\sum_{y\in M} e\{-B(y,x)\}\,e_y.\end{aligned} \tag{2.10}$$

Recall from Definition 1.8 that we have $\sigma(\underline{M}) = \frac{1}{\sqrt{|M|}}\sum_{x\in M} e\{-Q(x)\}$.

As it will turn out later we can extend the map $A \mapsto U(A)$ (A one of our generators) to a projective representation of Γ (Theorem 2.8). Hence we can find a central extension $\Gamma_{\underline{M}}$ of Γ which acts on $\mathbb{C}[M]$ such that the action of suitable preimages of S and T_b in the extension is given by the operators $U(S)$ and $U(T_b)$ (see Proposition 2.26). However, this extension would a priori depend on the particular underlying $\mathcal{O}$-FQM. Since, on the one hand side, we need such an extension (not necessarily central) which does not depend on the underlying $\mathcal{O}$-FQM, and since we do not want to analyze these extensions $\Gamma_{\underline{M}}$ more closely here, we adopt the following strategy.

For the rest of this chapter we fix once and for all a group $\tilde{\Gamma}$ such that the following four conditions are satisfied:

(i) $\tilde{\Gamma}$ acts on $W(\underline{M})$ for all $\mathcal{O}$-FQM $\underline{M}$
(ii) There exists T_b^* $(b \in \mathcal{O})$, $S^* \in \tilde{\Gamma}$ such that, for every $\mathcal{O}$-FQM $\underline{M}$, we have, for all $x \in \underline{M}$, the identities

$$T_b^* e_x = U(T_b) e_x, \quad S^* e_x = U(S) e_x, \tag{2.11}$$

where $U(T_b)$ and $U(S)$ are given by (2.10).
(iii) $\tilde{\Gamma}$ is generated by S^* and T_b^* $(b \in \mathcal{O})$.
(iv) There is an epimorphism from $\tilde{\Gamma}$ to Γ which maps S^* to S and T_b^* to T_b for b in $\mathcal{O}$.

In fact, such group exist. For example, we can take for $\tilde{\Gamma}$ the free group generated by elements S^* and T_b^* $(b \in \mathcal{O})$. In general, every group $\tilde{\Gamma}$ satisfying (2.11) will be a quotient of this free group. If K is totally real and if we restrict our theory to $\mathcal{O}$-FQM which are discriminant modules of a totally positive definite even $\mathcal{O}$-lattice (see Definition 3.3), then by Theorem 3.4 we can take for $\tilde{\Gamma}$ the metaplectic cover of $\mathrm{SL}(2, \mathcal{O})$ defined in the next chapter in Sect. 3.3.

Definition 2.27 Let $\underline{M}$ be an $\mathcal{O}$-FQM. We write $W(\underline{M})$ for the $\tilde{\Gamma}$-module $\mathbb{C}[M]$ with the $\tilde{\Gamma}$-action (2.11). By slight abuse of language, we shall refer to $W(\underline{M})$ as the *Weil representation associated to* $\underline{M}$. The Weil representation associated to an $\mathcal{O}$-CM is called a *cyclic Weil representation*.

As we stated in the beginning the goal of this chapter will be to decompose $W(\underline{M})$ into smaller $\tilde{\Gamma}$-submodules. For this the particular choice of $\tilde{\Gamma}$ is not important as it is explained by the following proposition, whose obvious proof we leave to the reader.

Proposition 2.28 *Suppose $\tilde{\Gamma}_1$ is a group satisfying (i), (ii), (iii) and (iv) (with Γ replaced by $\tilde{\Gamma}_1$). If $W(\underline{M}) = \bigoplus_j M_j$, where the M_j are $\tilde{\Gamma}$-submodules, then the M_j are also $\tilde{\Gamma}_1$-submodules. If $M_j \subseteq W(\underline{M})$ is irreducible with respect to $\tilde{\Gamma}$, then it is also irreducible with respect to $\tilde{\Gamma}_1$.*

It is, of course, still an interesting question to describe the smallest quotient of the free group generated by elements S^* and T_b^* $(b \in \mathcal{O})$ which we can take for $\tilde{\Gamma}$. In fact, the precise answer can be given. Since we do not need it here we confine ourselves to describe it at the end of this section. Before doing so we note an important property of the Weil representations, which will be needed later.

By $\langle \_|\_ \rangle$, we denote the Hermitian scalar product on $W(\underline{M})$ which is anti-linear in the second argument and which satisfies:

$$\langle e_x \mid e_y \rangle = \begin{cases} 1 & \text{if } x = y \\ 0 & \text{otherwise.} \end{cases} \tag{2.12}$$

Proposition 2.29 *Let $\underline{M} = (M, Q)$ be an $\mathcal{O}$-FQM. The operators $U(T_b)$ $(b \in \mathcal{O})$ and $U(S)$ are unitary with respect to the scalar product in* (2.12)*. In particular, the action of $\tilde{\Gamma}$ is unitary with respect to this scalar product.*

Proof It suffices to show that the operators $U(T_b)$ and $U(S)$ are unitary. For the former ones this is obvious. For proving that $U(S)$ is unitary let B be the bilinear form of $\underline{M}$ and let $v, v' \in W(\underline{M})$, so that $v = \sum_{x \in \underline{M}} v(x) e_x$ and $v' = \sum_{x' \in \underline{M}} v'(x') e_{x'}$. By (2.10) we have

$$U(S)v = \frac{\sigma(\underline{M})}{\sqrt{|M|}} \sum_{x \in M} v(x) \sum_{y \in M} e\{-B(y,x)\}\, e_y,$$

and similarly for v'. Hence we have

$$\langle U(S)v | U(S)v' \rangle = \frac{|\sigma(\underline{M})|^2}{|M|} \sum_{x,x' \in M} v(x)\overline{v'(x')} \sum_{y \in M} e\left\{B(y, -x + x')\right\}.$$

The inner sum equals zero if $x' \neq x$ (see Proposition 1.11), and otherwise it equals $|M|$. From Proposition 1.10, we know $|\sigma(M)| = 1$. It follows that $U(S)$ is unitary. □

We describe the smallest group $\tilde{\Gamma}$ satisfying (2.11). For this we describe, first of all, the smallest central extensions of $\mathrm{SL}(2, \mathcal{O})$ which the projective representations (2.10) can be lifted to.

We use $\kappa_{\mathfrak{p}}$ for the Kubota symbol at the prime $\mathfrak{p}$, i.e. for the $\mathrm{SL}(2, K_{\mathfrak{p}})$ 2-cocycle

$$\kappa_{\mathfrak{p}}(A, B) = \left(\frac{x(A)}{x(AB)}, \frac{x(B)}{x(AB)} \right)_{\mathfrak{p}}.$$

Here $(\_,\_)_{\mathfrak{p}}$ denotes the Hilbert symbol of the completion $K_{\mathfrak{p}}$ of the field K at the prime $\mathfrak{p}$, and where, for $A = \left(\begin{smallmatrix} a & b \\ c & d \end{smallmatrix}\right)$, we use $x(A) = c$ if $c \neq 0$ and $x(A) = d$ otherwise. The Kubota symbol defines a 2-cocycle of $\mathrm{SL}(2, K_{\mathfrak{p}})$ [Kub67, Thm.]. For a finite set S of primes of K, we use $\mathrm{Mp}(2, \mathcal{O}, S)$ for the central extension of $\mathrm{SL}(2, \mathcal{O})$ by the group $\{\pm 1\}$ which consists of all pairs (A, t) with A on $\mathrm{SL}(2, \mathcal{O})$ and $t = \pm 1$, equipped with the multiplication

$$(A, t) \cdot (B, t') = \Big(AB, tt' \prod_{\mathfrak{p} \in S} \kappa_{\mathfrak{p}}(A, B)\Big).$$

For a given prime ideal $\mathfrak{p}$ and $\mathfrak{p}$-module $\underline{M}$, we define a function on the group of units $\mathscr{O}^*$ of $\mathscr{O}$ by setting

$$\varepsilon_{\underline{M}}(a) = \sigma(M^{-a})/\sigma(M^{-1}).$$

One can prove that either $\varepsilon_{\underline{M}}(a)$ is a character of $\mathscr{O}^*$ for odd $\mathfrak{p}$, whereas this is not always true if $\mathfrak{p}$ is even (see [BS14a, Prop. 5.1]). For a given $\mathscr{O}$-FQM let

$$S_{\underline{M}} := \{\mathfrak{p} : \varepsilon_{\underline{M}(\mathfrak{p})} \text{ is not a character}\}.$$

Thus $S_{\underline{M}}$ is a subset of the prime ideals of K dividing 2.

Theorem 2.2 ([BS14a, Thm. 6.2]) *For a given $\mathscr{O}$-FQM $\underline{M} = (M, Q)$, the applications $(T_b, t) \mapsto tU(T_b)$ and $(S, t) \mapsto tU(S)$ can be extended to a representation $U : \mathrm{Mp}(2, \mathscr{O}, S_{\underline{M}}) \to \mathrm{GL}(\mathbb{C}[M])$.*

Finally, the smallest $\tilde{\Gamma}$ satisfying (2.11) is the projective limit of the projective system built from the finitely many groups $\mathrm{Mp}(2, \mathscr{O}, S_{\underline{M}})$ together with the obvious homomorphisms from $\mathrm{Mp}(2, \mathscr{O}, S_{\underline{M}})$ onto $\mathrm{Mp}(2, \mathscr{O}, S'_{\underline{M}})$ for $S_{\underline{M}} \supseteq S'_{\underline{M}}$. More precisely, one can describe $\tilde{\Gamma}$ as the group of all elements (A, t), where t is an element of $\{\pm 1\}^n$, and where the multiplication is given by

$$\left(A, \{t_{\mathfrak{p}}\}_{\mathfrak{p}|2}\right) \cdot \left(B, \{t'_{\mathfrak{p}}\}_{\mathfrak{p}|2}\right) = \left(AB, \{t_{\mathfrak{p}} \cdot t'_{\mathfrak{p}} \cdot \kappa_{\mathfrak{p}}(A, b)\}_{\mathfrak{p}|2}\right).$$

Each $\mathrm{Mp}(2, \mathscr{O}, S_{\underline{M}})$ is a quotient of $\tilde{\Gamma}$, namely, as homomorphic image under the map

$$\tilde{\Gamma} \to \mathrm{Mp}(2, \mathscr{O}, S_{\underline{M}}), \quad \left(A, \{t_{\mathfrak{p}}\}_{\mathfrak{p}|2}\right) \mapsto \left(A, \prod_{\mathfrak{p} \in S_{\underline{M}}} t_{\mathfrak{p}}\right).$$

2.3 Decomposition of Weil Representations

The purpose of this section will be to determine subrepresentations of the $\tilde{\Gamma}$-modules $W(\underline{M})$, which were introduced in the preceding section. We shall not derive a complete decomposition in general. However our results will suffice to give a complete decomposition of $W(\underline{M})$ in the important case of a cyclic $\underline{M}$. The main results are Theorems 2.3 and 2.4.

For our language concerning representations of groups the reader is referred to Sect. 2.1. For the notions concerning finite quadratic $\mathscr{O}$-modules that we use in the sequel, e.g. orthogonal groups, isotropic modules an so on, we refer to reader to Sect. 1.1.

We shall first explain the main results of this section and state three auxiliary propositions which are needed for the understanding of the main results. The rest of this section is dedicated to the proofs of these auxiliary propositions (Propositions 2.30, 2.34, and 2.35) and the proofs of the main results.

The decomposition of the $\tilde{\Gamma}$-modules is based on two principles. The first one is that the Weil representation of a quotient of an $\mathcal{O}$-FQM $\underline{M}$ embeds naturally into $W(\underline{M})$ as $\tilde{\Gamma}$-submodule.

Proposition 2.30 *Let $\underline{M} = (M, Q)$ be an $\mathcal{O}$-FQM and U be an isotropic submodule of $\underline{M}$. The linear map*

$$\iota_U : W(\underline{M}/U) \hookrightarrow W(\underline{M}), \quad e_X \mapsto \sum_{y\in X} e_y$$

defines a $\tilde{\Gamma}$-linear embedding (i.e. an injective $\tilde{\Gamma}$-module homomorphism).

Definition 2.31 Let $\underline{M} = (M, Q)$ be an $\mathcal{O}$-FQM. We define the *new part* $W(\underline{M})^{\text{new}}$ of $W(\underline{M})$ as the orthogonal complement of the subspace

$$\sum_{\substack{U\subseteq \underline{M}\\ U\,\text{isotropic}\\ U\neq 0}} \iota_U W(\underline{M}/U)$$

with respect to the scalar product in (2.12). Here the sum is over all isotropic submodules $U \neq 0$ of $\underline{M}$.

Remark By Proposition 2.30 the spaces $\iota_U W(\underline{M}/U)$ are $\tilde{\Gamma}$-invariant, and hence their sum is so too. Since $\tilde{\Gamma}$ acts unitarily (see Proposition 2.29) the space $W(\underline{M})^{\text{new}}$ is in fact a $\tilde{\Gamma}$-submodule of $W(\underline{M})$.

Theorem 2.3 *Let $\underline{M} = (M, Q)$ be an $\mathcal{O}$-FQM. We have the following decomposition of $W(\underline{M})$ into $\tilde{\Gamma}$-submodules:*

$$W(\underline{M}) = W(\underline{M})^{new} \oplus \sum_{\substack{U\subseteq \underline{M}\\ U\ isotropic\\ U\neq 0}} \iota_U W(\underline{M}/U)^{new}. \tag{2.13}$$

If $\underline{M}$ contains only one maximal isotropic submodule, then the second sum in (2.13) *is an orthogonal sum with respect to the scalar product* (2.12).

Remark Note that the decomposition (2.13) is a direct sum decomposition for $\mathcal{O}$-CM, since a cyclic $\underline{M}$ contains only one maximal isotropic submodule (see the remark after Theorem 1.2). Recall also that there exist also $\mathcal{O}$-FQM which are not cyclic but contain only one maximal isotropic submodule. The condition that there exists only one maximal isotropic submodule is not necessary for the decomposition in (2.13) to be direct as the subsequent Example 2.32 shows. However, this condition is also not superfluous as we shall show in the Example 2.33 below.

Example 2.32 We show that the sum (2.13) applied to the finite quadratic $\mathbb{Z}$-module $\underline{N} := (\mathbb{Z}/2\mathbb{Z} \times \mathbb{Z}/2\mathbb{Z}, Q)$, where $Q(x + 2\mathbb{Z}, y + 2\mathbb{Z}) = xy/2 + \mathbb{Z}$, is direct. The nonzero isotropic submodules of $\underline{N}$ are $U_1 = \langle([0],[1])\rangle$, $U_2 = \langle([1],[0])\rangle$. (Here we use $[x] = x + 2\mathbb{Z}$.) Since $|U_i^{\#}| \cdot |U_i| = 4$ (Proposition 1.7) the quotient modules $\underline{N}/U_i$ are trivial, in particular, $W(\underline{N}/U_i) = W(\underline{N}/U_i)^{\text{new}}$. They are spanned by the vectors $e_{([0],[0])} + e_{([0],[1])}$ and $e_{([0],[0])} + e_{([1],[0])}$, respectively, which are obviously linearly independent. We thus have $W(\underline{N}) = W(\underline{N})^{\text{new}} \oplus \iota_{U_1} W(\underline{N}/U_1)^{\text{new}} \oplus \iota_{U_2} W(\underline{N}/U_2)^{\text{new}}$.

Example 2.33 Let $\underline{N'} := (\mathbb{Z}/2\mathbb{Z} \times \mathbb{Z}/2\mathbb{Z}, Q')$, where Q' denotes the quadratic form $Q'(x + 2\mathbb{Z}, y + 2\mathbb{Z}) = (x^2 + xy + y^2)/2 + \mathbb{Z}$. We show that the sum (2.13) applied to $\underline{M} := \underline{N'} \oplus \underline{N}$, where $\underline{N}$ is as in Example 2.32, is not direct. The nonzero isotropic submodules of $\underline{M}$ are $U_1 = \big\langle([0],[0]) \oplus ([0],[1])\big\rangle$, $U_2 = \big\langle([0],[0]) \oplus ([1],[0])\big\rangle$, $U_3 = \big\langle([1],[1]) \oplus ([1],[1])\big\rangle$, $U_4 = \big\langle([0],[1]) \oplus ([1],[1])\big\rangle$ and $U_5 = \big\langle([1],[0]) \oplus ([1],[1])\big\rangle$. Note that, for all i, U_i is maximal. The order of $\underline{M}/U_i$ equals 4 (Proposition 1.7). Since the U_i are maximal, the $\mathbb{Z}$-FQM $\underline{M}/U_i$ are anisotropic, i.e have no nonzero isotropic submodules. (In fact, one can show that $\underline{M}/U_i$ is isomorphic to $\underline{N'}$.) Hence we have $\iota_{U_i} W(\underline{M}/U_i) = \iota_{U_i} W(\underline{M}/U_i)^{new}$. Since $W(\underline{M})$ has dimension 16 the sum of the five four-dimensional spaces $\iota_{U_i} W(\underline{M}/U_i)^{new}$ cannot be direct.

The second principle for decomposing a Weil representation $W(\underline{M})$ is the natural action of the orthogonal group $\mathrm{O}(\underline{M})$ on $W(\underline{M})$ coming from the permutation representations given by its action on $\underline{M}$. The main observation is that this action intertwines with the action of $\tilde{\Gamma}$.

Proposition 2.34 *The group $\mathrm{O}(\underline{M})$ acts on the space $W(\underline{M})$ via linear continuation of the map:*

$$\big(\varphi, e_x\big) \mapsto \varphi\, e_x := e_{\varphi(x)}. \tag{2.14}$$

This action is unitary with respect to the scalar product 2.12. The action of $\mathrm{O}(\underline{M})$ and the action of $\tilde{\Gamma}$ on $W(\underline{M})$ commute.

In fact, the action of the orthogonal groups enables us to further decompose the spaces $W(\underline{M}))^{\text{new}}$ into $\tilde{\Gamma}$-submodules as is explained by the next proposition.

Proposition 2.35 *Let $\underline{M} = (M, Q)$ be an $\mathcal{O}$-FQM. The space $W(\underline{M})^{new}$ is $\mathrm{O}(\underline{M})$-invariant.*

Theorem 2.4 *Let $\underline{M} = (M, Q)$ be an $\mathcal{O}$-FQM. For each irreducible character of $\mathrm{O}(\underline{M})$, the sum $W(\underline{M})^{new,\chi}$ of those $\mathrm{O}(\underline{M})$-submodules of $W(\underline{M})^{new}$ which afford the character χ, is invariant under $\tilde{\Gamma}$. In particular, we have the decomposition of $W(\underline{M})^{new}$ into $\tilde{\Gamma}$-submodules*

$$W(\underline{M})^{new} = \bigoplus_{\chi \in \widehat{\mathrm{O}(\underline{M})}} W(\underline{M})^{new,\chi}. \tag{2.15}$$

(Recall that $\widehat{\mathrm{O}(\underline{M})}$ *denotes the set of irreducible characters of the orthogonal group* $\mathrm{O}(\underline{M})$*.)*

Remark Note that the components of the decomposition (2.15) are in general not irreducible $\tilde{\Gamma}$-modules. However, for $\mathscr{O}$-CM, they turn out be irreducible (see Sect. 2.4).

Proof of Proposition 2.30 Let B stand for the bilinear form associated of $\underline{M}$. It is enough to prove the result for the elements T_b^* ($b \in \mathscr{O}$), S^*. Let $b \in \mathscr{O}$. Using the T_b^*-action in (2.11), the result holds true for T_b^*, since we have the following identity for any $x \in U^\#$:

$$\begin{aligned}\iota_U(T_b^* e_{x+U}) &= e\{bQ(x)\}\,\iota_U(e_{x+U}) = e\{bQ(x)\} \sum_{y \in x+U} e_y \\ &= \sum_{y \in x+U} e\{bQ(y)\}\, e_y = \sum_{y \in x+U} T_b^* e_y = T_b^* \iota_U(e_{x+U}).\end{aligned}$$

The third identity follows from the very definition of isotropic submodules.

We set $C_{\underline{M}/U} := \sigma(\underline{M}/U)\frac{1}{\sqrt{|U^\#/U|}}$ and $C_{\underline{M}} := \sigma(\underline{M})\frac{1}{\sqrt{|M|}}$. To prove the claimed identity for S^*, first we determine $\iota_U(S^* e_{x+U})$. Later we shall compare this with $S^* \iota_U(e_{x+U})$. For any $x \in U^\#$, using the S^*-action in (2.11), we have

$$\begin{aligned}\iota_U(S^* e_{x+U}) &= C_{\underline{M}/U} \sum_{y+U \in \underline{M}/U} e\{-\underline{B}(y+U, x+U)\}\, \iota_U e_{y+U} \\ &= C_{\underline{M}/U} \sum_{y+U \in \underline{M}/U} e\{-\underline{B}(y+U, x+U)\} \sum_{y' \in y+U} e_{y'} \\ &= C_{\underline{M}/U} \sum_{y+U \in \underline{M}/U} \sum_{y' \in y+U} e\{-\underline{B}(y+U, x+U)\}\, e_{y'} \\ &= C_{\underline{M}/U} \sum_{y \in U^\#} e\{-B(y, x)\}\, e_y.\end{aligned}$$

On the other hand, again from the S^*-action in (2.11), we have

$$\begin{aligned}S^* \iota_U(e_{x+U}) &= \sum_{y' \in x+U} S^* e_{y'} = C_{\underline{M}} \sum_{y' \in x+U} \sum_{y \in \underline{M}} e\{-B(y, y')\}\, e_y \\ &= C_{\underline{M}} \sum_{y \in \underline{M}} e_y \sum_{y' \in x+U} e\{-B(y', y)\} \\ &= C_{\underline{M}} \sum_{y \in \underline{M}} e\{-B(y, x\}\, e_y \sum_{u \in U} e\{-B(y, u)\} \\ &= |U| C_{\underline{M}} \sum_{y \in U^\#} e\{-B(y, x)\}\, e_y.\end{aligned}$$

For the last identity we used the fact that the inner sum in the previous sum is zero unless $y \in U^\#$, when it equals $|U|$. To obtain the claimed identity for S^*, it remains to prove the identity $|U|C_{\underline{M}} = C_{\underline{M}/U}$. But from Proposition 1.9, we have $\sigma(\underline{M}) = \sigma(\underline{M}/U)$ and from Proposition 1.7, we have $\sqrt{|M|} = |U|\sqrt{|U^\#/U|}$. □

Remark For every isotropic submodule U of $\underline{M}$, the orthogonal projection of $W(\underline{M})$ onto $\iota_U W(\underline{M}/U)$ is given by the formula:

$$P_U^{\underline{M}}(v) = \frac{1}{\sqrt{|U|}} \sum_{X \in \underline{M}/U} \left(\sum_{x \in X} v(x) \right) \sum_{x \in X} e_x.$$

This follows from the fact that the vectors $\frac{1}{\sqrt{|U|}} \sum_{x \in X} e_x$ $(X \in \underline{M}/U)$ form an orthonormal basis of the space $\iota_U W(\underline{M}/U)$. Note that v is in $W(\underline{M})^{\text{new}}$ if and only if $P_U^{\underline{M}}(v) = 0$ for all $U \neq 0$.

For the proof of Theorem 2.3, we need two lemmas.

Lemma 2.36 *Let $\underline{M}$ be an $\mathcal{O}$-FQM and $U \subseteq V$ be isotropic submodules of $\underline{M}$. The following diagram is commutative*

$$\begin{array}{ccc} W(\underline{M}/U/V/U) & \xrightarrow{\varphi} & W(\underline{M}/V) \\ \downarrow{\scriptstyle \iota_{V/U}} & & \downarrow{\scriptstyle \iota_V} \\ W(\underline{M}/U) & \xrightarrow{\iota_U} & W(\underline{M}), \end{array}$$

where φ is induced by the canonical isomorphism $(x+U)+V/U \mapsto x+V$.

Proof First note that $V \subseteq U^\#$, since V isotropic (i.e. $V \subseteq V^\#$) and $V^\# \subseteq U^\#$ (see the assumption). Note also that V/U is an isotropic submodule of $\underline{M}/U$ and $(V/U)^\#$ equals $V^\#/U$. This shows that the map φ is well-defined.

The following identity proves the lemma:

$$\iota_V \circ \varphi(e_{(x+U)+V/U}) = \iota_V(e_{x+V}) = \sum_{y \in x+V} e_y = \sum_{y+U \in (x+U)+V/U} \sum_{y' \in y+U} e_{y'}$$

$$= \sum_{y+U \in (x+U)+V/U} \iota_U(e_{y+U}) = \iota_U \circ \iota_{V/U}(e_{(x+U)+V/U}).$$

□

Lemma 2.37 *Let U, V be isotropic submodules of the $\mathcal{O}$-FQM $\underline{M}$ such that $U+V$ is isotropic. Then we have*

$$P_U^{\underline{M}} \iota_V = \sqrt{|U \cap V|}\, \iota_V P_{(U+V)/V}^{\underline{M}/V}. \tag{2.16}$$

Proof By the remark after Theorem 2.3, we have

$$P_U^M(e_z) = \frac{1}{\sqrt{|U|}} \sum_{\substack{y \in U^\# \\ y \equiv z \bmod U}} e_y \quad (z \in M).$$

First we evaluate the left hand side of (2.16) at e_{x_0+V} $(x_0 \in V^\#)$. We have

$$\begin{aligned} P_U^M \iota_V(e_{x_0+V}) &= \frac{1}{\sqrt{|U|}} \sum_{y \in x_0+V} \sum_{\substack{z \in U^\# \\ z \equiv y \bmod U}} e_z = \frac{1}{\sqrt{|U|}} \sum_{v \in V} \sum_{\substack{z \in U^\# \\ z \equiv x_0+v \bmod U}} e_z \\ &= \frac{1}{\sqrt{|U|}} \sum_{v \in V} \sum_{\substack{u \in U \\ x_0+v+u \in U^\#}} e_{x_0+v+u}. \end{aligned}$$

The map

$$\{(u, v) \in U + V : x_0 + v + u \in U^\#\} \stackrel{\varphi}{\to} \{z \in U^\# \cap V^\# : z \equiv x_0 \bmod U + V\}$$

given by $\varphi(u, v) = x_0 + v + u$ is obviously a surjective map. The well-definedness of φ follows from the fact that $U \subseteq V^\#$ and $V \subseteq U^\#$, since $U + V$ is isotropic. We claim that each fiber has $|U \cap V|$-many elements. Let z be an element of the right hand side. Since φ is surjective, there exist (u, v) such that $\varphi(u, v) = z$. The number of elements in the fiber of z equals the number of elements in $\{(u, v) \in U + V : u + v \equiv 0 \bmod U + V\}$. But this set has clearly $|U \cap V|$-many elements. Therefore, we have

$$P_U^M \iota_V(e_{x_0+V}) = |U \cap V| \frac{1}{\sqrt{|U|}} \sum_{\substack{z \in U^\# \cap V^\# \\ z \equiv x_0 \bmod U+V}} e_z. \tag{2.17}$$

Since $\left((U + V)/V\right)^\# = \left(U^\# \cap V^\#\right)/V$, we have

$$\begin{aligned} \iota_V P_{(U+V)/V}^{M/V}(e_{x_0+V}) &= \frac{1}{\sqrt{|(U + V)/V|}} \sum_{\substack{Y \in (U^\# \cap V^\#)/V \\ Y \equiv x_0+V \bmod (U+V)/V}} \iota_V e_Y \\ &= \frac{1}{\sqrt{|(U + V)/V|}} \sum_{\substack{Y \in (U^\# \cap V^\#)/V \\ Y \equiv x_0+V \bmod (U+V)/V}} \sum_{y \in Y} e_y. \end{aligned}$$

By doing the substitution $Y \mapsto \pi(y)$ (where $Y = y + V$ and π is the canonical projection from $U^{\#} \cap V^{\#}$ onto $\big(U^{\#} \cap V^{\#}\big)/V$), we obtain

$$\iota_V P^{\underline{M}/V}_{(U+V)/V}(e_{x_0+V}) = \frac{1}{|V|}\frac{1}{\sqrt{|(U+V)/V|}} \sum_{\substack{y \in U^{\#} \cap V^{\#} \\ y \equiv x_0 \bmod U+V}} \sum_{y' \equiv y \bmod V} e_{y'}.$$

We study the map φ' from $\{(y, y') \in (U^{\#} \cap V^{\#})^2 : y \equiv x_0 \bmod U + V, y' \equiv y \bmod V\}$ to $\{z \in U^{\#} \cap V^{\#} : z \equiv x_0 \bmod U + V\}$ which is defined by $(y, y') \mapsto y'$. The map φ' is surjective. Indeed, let z be an element of the latter set. Then clearly $\varphi'(z, z) = z$. We claim that each fiber has $|V|$-many elements. The fiber of the element z equals $\{y \in U^{\#} \cap V^{\#} : y \equiv x_0 \bmod U + V, y \equiv z \bmod V\}$. But since $y - z \in V$ implies that $y - x_0 \bmod U + V$, we observe that the fiber of z has $|V|$-many elements. Therefore, we have

$$\iota_V P^{\underline{M}/V}_{(U+V)/V}(e_{x_0+V}) = \frac{1}{\sqrt{|(U+V)/V|}} \sum_{\substack{z \in U^{\#} \cap V^{\#} \\ z \equiv x_0 \bmod U+V}} e_z. \tag{2.18}$$

In view of the identities (2.17) and (2.18), to show that (2.16) holds true, it remains to prove the following identity:

$$|U \cap V|\frac{1}{\sqrt{|U|}} = \sqrt{|U \cap V|}\frac{1}{\sqrt{|(U+V)/V|}}.$$

However, from the second isomorphism theorem for modules we have the isomorphism $\big(U + V\big)/V \simeq U/\big(U \cap V\big)$, and hence the claimed identity holds true. □

Proof of Theorem 2.3 We proceed by induction on the order of M. If $\underline{M}$ does not possess isotropic submodules, then there is nothing to prove. Otherwise, by the definition of $W(\underline{M})^{\text{new}}$, we have

$$W(\underline{M}) = W(\underline{M})^{\text{new}} \oplus \sum_{\substack{U \subseteq \underline{M} \\ U \text{is isotropic} \\ U \neq 0}} \iota_U W(\underline{M}/U).$$

By induction hypothesis for $U \neq 0$, we can write

$$W(\underline{M}/U) = \sum_{\substack{V/U \subseteq \underline{M}/U \\ V/U \text{isotropic}}} \iota_{V/U} W\big((\underline{M}/U)/(V/U)\big)^{\text{new}}.$$

Inserting this into the first identity, we obtain

$$W(\underline{M}) = W(\underline{M})^{\text{new}} \oplus \sum_{\substack{U \subseteq \underline{M} \\ U \text{ is isotropic} \\ U \neq 0}} \sum_{\substack{V/U \subseteq \underline{M}/U \\ V/U \text{ isotropic}}} \iota_U \iota_{V/U} W\big((\underline{M}/U)/(V/U)\big)^{\text{new}}.$$

The claimed decomposition follows now by the identity

$$\iota_U \iota_{V/U} W\big((\underline{M}/U)/(V/U)\big)^{\text{new}} = \iota_V W(\underline{M}/V)^{\text{new}}$$

which is obvious from Lemma 2.36.

For proving the second statement of the theorem, assume that there is only one maximal isotropic submodule in $\underline{M}$, or, equivalently, that the set of isotropic submodules of $\underline{M}$ is closed under addition. Let U and V be isotropic submodules, $U \neq V$. It suffices to show that $\iota_V W(\underline{M}/V)^{\text{new}}$ is orthogonal to $\iota_U W(\underline{M}/U)$. By Lemma 2.37, we have

$$P_U^{\underline{M}} \iota_V W(\underline{M}/V)^{\text{new}} = \sqrt{|U \cap V|}\, \iota_V P_{(U+V)/V}^{\underline{M}/V} W(\underline{M}/V)^{\text{new}}.$$

Since $U \neq V$, we have $\big(U + V\big)/V \neq 0$. Hence the right hand side of the last identity is zero (see the second remark after the proof of Proposition 2.30). But this means $\iota_V W(\underline{M}/V)^{\text{new}}$ is in the kernel of the orthogonal projection $P_U^{\underline{M}}$, hence it is perpendicular to the image of $P_U^{\underline{M}}$, which equals $\iota_U W(\underline{M}/U)$. This proves the theorem. □

Proof of Proposition 2.34 It is clear that the map in the statement of the proposition defines indeed an action. The action is unitary with respect to (2.12), since the elements of the orthogonal group are in fact automorphisms on M.

We show in the following that the actions of $\mathrm{O}(\underline{M})$ and $W(\underline{M})$ commute. Let B be associated bilinear form of $\underline{M}$. Let $\varphi \in \mathrm{O}(\underline{M})$ and $b \in \mathcal{O}$, $x \in M$. The action of T_b^* (see (2.11)) and $\mathrm{O}(\underline{M})$ commute, since:

$$\begin{aligned}\varphi T_b^* e_x &= e\{bQ(x)\}\, \varphi e_x = e\{bQ(x)\}\, e_{\varphi(x)} = e\{bQ(\varphi(x))\}\, e_{\varphi(x)} \\ &= T_b^* e_{\varphi(x)} = T_b^* \varphi e_x.\end{aligned}$$

Similarly, the action of S^* (see (2.11)) and $\mathrm{O}(\underline{M})$ commute, since we have:

$$\begin{aligned}\varphi S^* e_x &= \frac{\sigma(M)}{\sqrt{|M|}} \sum_{y \in M} e\{-B(y,x)\}\, \varphi e_y = \frac{\sigma(M)}{\sqrt{|M|}} \sum_{y \in \underline{M}} e\{-B(y,x)\}\, e_{\varphi(y)} \\ &= \frac{\sigma(M)}{\sqrt{|M|}} \sum_{y \in \underline{M}} e\{-B(\varphi^{-1}(y),x)\}\, e_y = \frac{\sigma(M)}{\sqrt{|M|}} \sum_{y \in M} e\{-B(y,\varphi(x))\}\, e_y \\ &= S^* e_{\varphi(x)} = S^* \varphi e_x.\end{aligned}$$

To obtain the third identity, we did the substitution $\varphi(y) \mapsto y$ in the previous sum. The forth identity holds true by the very definition of the orthogonal group. □

For the proof of Proposition 2.35 we need a lemma.

Lemma 2.38 *Let $\underline{M} = (M, Q)$ be an $\mathcal{O}$-FQM and U be an isotropic submodule of $\underline{M}$. If U is fixed by $\mathrm{O}(\underline{M})$, then for $\varphi \in \mathrm{O}(\underline{M})$, we have*

$$\varphi\, \iota_U = \iota_U \phi_U(\varphi), \tag{2.19}$$

where $\phi_U : \mathrm{O}(\underline{M}) \to \mathrm{O}(\underline{M}/U)$ is defined by $\phi_U(\varphi)(x + U) = \varphi(x) + U$.

Proof Let B be associated bilinear form on $\underline{M}$. We need to show first of all that the map ϕ_U is well-defined. Let $\varphi \in \mathrm{O}(\underline{M})$. First we show that for any $x \in U^{\#}$, $\varphi(x)$ is an element of $U^{\#}$, i.e. $B(\varphi(x), U) = 0$. But this follows from the very definition of the orthogonal group. By the same reasoning, we also have $\phi_U(\varphi) \in \mathrm{O}(\underline{M}/U)$. For the well-definedness it remains to show that ϕ_U does not depend on the choice of the representatives of $U^{\#}/U$. Let $x' \in x + U$. Write $x' = x + u$ ($u \in U$). But we have $\varphi(x') - \varphi(x) = \varphi(x + u) - \varphi(x) = \varphi(u) \in U$. The last identity follows from the assumption that $\varphi(U) = U$.

The statement of the lemma holds true, since we have:

$$\begin{aligned}\varphi\, \iota_U(e_{x+U}) &= \sum_{y \in x+U} \varphi e_y = \sum_{y \in x+U} e_{\varphi(y)} = \sum_{\varphi^{-1}(y) \in x+U} e_y \\ &= \sum_{y \in \varphi(x)+U} e_y = \iota_U(e_{\varphi(x)+U}) = \iota_U(\phi_U(\varphi) e_{x+U}).\end{aligned}$$

For the third identity we did the substitution $\varphi(y) \mapsto y$ in the previous sum. To obtain the forth identity we used the assumption $\varphi(U) = U$. □

Proof of Proposition 2.35 This follows immediately from Proposition 2.34 and Lemma 2.38. □

For the proof of Theorem 2.4, we need a lemma.

Lemma 2.39 *Let $\underline{M}$ be an $\mathcal{O}$-FQM. The space $W(\underline{M})^{new,\chi}$ ($\chi \in \widehat{\mathrm{O}(\underline{M})}$) is a $\tilde{\Gamma}$-submodule of $W(\underline{M})^{new}$.*

Proof Write $W(\underline{M})^{\mathrm{new},\chi} = \sum_{i \in I} W_i$, where $\{W_i\}_{i \in I}$ is the set of all $\mathrm{O}(\underline{M})$-submodules of $W(\underline{M})^{\mathrm{new}}$ affording the character χ. It suffices to show that the spaces αW_i for $\alpha \in \tilde{\Gamma}$, is again an $\mathrm{O}(\underline{M})$-submodule of $W(\underline{M})^{\mathrm{new}}$ affording the character χ. But this follows immediately from the fact that $x \mapsto \alpha x$ defines an $\mathrm{O}(\underline{M})$-module isomorphism of W_i and αW_i since the actions of $\tilde{\Gamma}$ and of $\mathrm{O}(\underline{M})$ commute as Proposition 2.34 shows. □

Proof of Theorem 2.4 Proposition 2.35 implies that the space $W(\underline{M})^{\mathrm{new}}$ is $\mathrm{O}(\underline{M})$-invariant. Hence, by Proposition 2.17, we have the decomposition as stated in the theorem. Finally by Lemma 2.39, we know that the components of the decomposition are $\tilde{\Gamma}$-submodules of $W(\underline{M})^{\text{new}}$. □

2.4 Complete Decomposition of Cyclic Representations

In this section we shall show that, for a cyclic $\mathcal{O}$-module $\underline{M}$, the decomposition of $W(\underline{M})$ resulting from the combination of Theorems 2.3 and 2.4 is complete, i.e. the components occurring in the decomposition are all irreducible. In addition, we shall derive dimension formulas for these irreducible components.

Recall from Sect. 1.2 that for an $\mathcal{O}$-FQM $\underline{M} = (M, Q)$ the level $\mathfrak{l}$, the modified level $\mathfrak{m}$ and the annihilator $\mathfrak{a}$ satisfy $\mathfrak{m} = \mathfrak{l}(2, \mathfrak{l})^{-2}$ and $\mathfrak{a} = \mathfrak{l}(2, \mathfrak{l})^{-1}$. Recall also, that, for cyclic $\underline{M}$, the isotropic submodules are all of the form $\mathfrak{a}\mathfrak{b}^{-1}M$, where $\mathfrak{b}$ runs through the square divisors of $\mathfrak{m}$. Finally recall, that, for a cyclic $\underline{M}$, the elements of the orthogonal group $\mathrm{O}(\underline{M})$ are given by multiplication by the elements g in the subgroup $\mathrm{E}(\underline{M}) \subseteq (\mathcal{O}/\mathfrak{a})^*$ of all $\varepsilon + \mathfrak{a}$ such that $\varepsilon^2 \equiv 1 \bmod \mathfrak{l}$ (see Proposition 1.21). Via this identification of $\mathrm{O}(\underline{M})$ with $\mathrm{E}(\underline{M})$ we shall henceforth consider $W(\underline{M})$ as an $\mathrm{E}(\underline{M})$-module via the action $(g, v) \mapsto gv$, where $(gv)(x) = v(\varepsilon x)$ if $g = \varepsilon + \mathfrak{a}$.

In the following we consider $W(\underline{M})$ as an $\mathrm{E}(\underline{M})$-module.

Definition 2.40 Let $\underline{M} = (M, Q)$ be an $\mathcal{O}$-CM with level $\mathfrak{l}$ and modified level $\mathfrak{m}$. For a square-free divisor $\mathfrak{f}$ of $\mathfrak{m}$, we set

$$W(\underline{M})^{\mathfrak{f}} := \{v \in W(\underline{M}) : v(gx) = \psi_{\mathfrak{f}}(g)\, v(x) \text{for all } g \in \mathrm{E}(\underline{M}),\, x \in M\}.$$

Here $\psi_{\mathfrak{f}}$ denotes the linear character $\psi_{\mathfrak{f}}(\varepsilon + \mathfrak{a}) = \mu\big(\mathfrak{f}, (\varepsilon + 1)(2, \mathfrak{l})^{-1}\big)$ of $\mathrm{E}(\underline{M})$ (see Proposition 1.23). Moreover, we define

$$W(\underline{M})^{\text{new},\mathfrak{f}} := W(\underline{M})^{\text{new}} \cap W(\underline{M})^{\mathfrak{f}}.$$

Remark Note that for an $\mathcal{O}$-CM $\underline{M}$, the spaces $W(\underline{M})^{\text{new},\mathfrak{f}}$ coincide with the spaces $W(\underline{M})^{\text{new},\psi'_{\mathfrak{f}}}$ occurring in Theorem 2.4, where $\psi'_{\mathfrak{f}}$ is the character of $\mathrm{O}(\underline{M})$ corresponding to $\psi_{\mathfrak{f}}$ under the isomorphism $\mathrm{E}(\underline{M}) \simeq \mathrm{O}(\underline{M})$ of Proposition 1.21.

Theorem 2.5 *Let* $\underline{M} = (M, Q)$ *be an* $\mathcal{O}$*-CM with level* $\mathfrak{l}$*, modified level* $\mathfrak{m}$ *and annihilator* $\mathfrak{a}$*.*

(i) We have the following decomposition of $W(\underline{M})$ *into* $\tilde{\Gamma}$*-submodules:*

$$W(\underline{M}) = \bigoplus_{\mathfrak{b}^2 \mid \mathfrak{m}} \iota_{\mathfrak{a}\mathfrak{b}^{-1}M} W(\underline{M}/\mathfrak{a}\mathfrak{b}^{-1}M)^{new}. \tag{2.20}$$

Here the sum is over all integral $\mathcal{O}$*-ideals* $\mathfrak{b}$ *whose square divides* $\mathfrak{m}$*.*

(ii) For $W(\underline{M})^{new}$ we have the decomposition

$$W(\underline{M})^{new} = \bigoplus_{\substack{\mathfrak{f}|\mathfrak{m} \\ \mathfrak{f}\, square-free}} W(\underline{M})^{new,\mathfrak{f}} \tag{2.21}$$

into $\tilde{\Gamma}$-submodules. The $W(\underline{M})^{new,\mathfrak{f}}$ are irreducible $\tilde{\Gamma}$-submodules.

(iii) For any square-free divisor $\mathfrak{f}$ of $\mathfrak{m}$, we have

$$\dim W(\underline{M})^{new,\mathfrak{f}} = \mathrm{N}(\mathfrak{a}) \prod_{\mathfrak{p}\|\mathfrak{m}} \frac{1}{2}\left(1 + \frac{\mu(\mathfrak{f},\mathfrak{p})}{\mathrm{N}(\mathfrak{p})}\right) \prod_{\mathfrak{p}^2|\mathfrak{m}} \frac{1}{2}\left(1 - \frac{1}{\mathrm{N}(\mathfrak{p}^2)}\right). \tag{2.22}$$

We subdivide the proof of the theorem into three parts.

Proof of Theorem 2.5 (i) The decomposition given in (i) is the decomposition of Theorem 2.3 specialized here to the case of cyclic finite quadratic $\mathcal{O}$-modules. The directness of the sum comes from the fact that a cyclic $\mathcal{O}$-FQM fulfills the assumption stated in Theorem 2.3, namely that it possesses only one maximal isotropic submodule (see the remark after Theorem 1.2). □

Next we shall prove (iii). For that we need a lemma.

Lemma 2.41 *Suppose $\underline{M} = (M, Q)$ be an $\mathcal{O}$-CM with annihilator $\mathfrak{a}$ and modified level $\mathfrak{m}$. The character $\chi_{W(\underline{M})^{new}}$ of the $\mathrm{E}(\underline{M})$-module $W(\underline{M})^{new}$ satisfies*

$$\chi_{W(\underline{M})^{new}}(\varepsilon + \mathfrak{a}) = \sum_{\mathfrak{b}^2|\mathfrak{m}} \mu(\mathfrak{b})\, \mathrm{N}\left(\varepsilon - 1, \mathfrak{a}\mathfrak{b}^{-2}\right).$$

Proof Since the space $W(\underline{M})$ is $\mathrm{O}(\underline{M})$-invariant (see (2.14)), using the decomposition in part (i) of Theorem 2.5 and also Proposition 2.10, we have

$$\mathrm{tr}\left(\varepsilon + \mathfrak{a}, W(\underline{M})\right) = \sum_{\mathfrak{b}^2|\mathfrak{m}} \mathrm{tr}\left(\varepsilon + \mathfrak{a}\mathfrak{b}^{-2}, W(\underline{M}/\mathfrak{a}\mathfrak{b}^{-1}M)^{\mathrm{new}}\right). \tag{2.23}$$

Here we also used the identity

$$\mathrm{tr}\left(\varepsilon + \mathfrak{a}, \iota_{\mathfrak{a}\mathfrak{b}^{-1}M} W(\underline{M}/\mathfrak{a}\mathfrak{b}^{-1}M)^{\mathrm{new}}\right) = \mathrm{tr}\left(\varepsilon + \mathfrak{a}\mathfrak{b}^{-2}, W(\underline{M}/\mathfrak{a}\mathfrak{b}^{-1}M)^{\mathrm{new}}\right)$$

which is a consequence of Lemma 2.38 (when we apply this lemma we used Corollary 1.19, which says that the annihilator of $\underline{M}/\mathfrak{a}\mathfrak{b}^{-1}M$ equals $\mathfrak{a}\mathfrak{b}^{-2}$). If we can show that the following identity holds true

$$\mathrm{tr}\left(\varepsilon + \mathfrak{a}, W(\underline{M})^{\mathrm{new}}\right) = \sum_{\mathfrak{b}^2|\mathfrak{m}} \mu(\mathfrak{b})\, \mathrm{tr}\left(\varepsilon + \mathfrak{a}\mathfrak{b}^{-2}, W(\underline{M}/\mathfrak{a}\mathfrak{b}^{-1}M)\right), \tag{2.24}$$

then the claimed formula holds true. Indeed, since $W(\underline{M})$ is a permutation representation with respect to the action of the group $\mathrm{O}(\underline{M})$ (see (2.14)), we then obviously have

$$\mathrm{tr}\left(\varepsilon + \mathfrak{a}\mathfrak{b}^{-2}, W(\underline{M}/\mathfrak{a}\mathfrak{b}^{-1}M)\right) = \mathrm{N}\left(\varepsilon - 1, \mathfrak{a}\mathfrak{b}^{-2}\right).$$

Now we prove (2.24). First we calculate the right hand side of (2.24). By inserting the value in (2.23) specialized to $\mathrm{tr}\left(\varepsilon + \mathfrak{a}\mathfrak{b}^{-2}, W(\underline{M}/\mathfrak{a}\mathfrak{b}^{-1}M)\right)$ and using Corollary 1.19 (which says that the annihilator and the modified level of $\underline{M}/\mathfrak{a}\mathfrak{b}^{-1}M$ equal $\mathfrak{a}\mathfrak{b}^{-2}$ and $\mathfrak{m}\mathfrak{b}^{-2}$, respectively), we obtain

$$\sum_{\mathfrak{b}^2|\mathfrak{m}} \mu(\mathfrak{b}) \sum_{\mathfrak{b}'^2|\mathfrak{m}\mathfrak{b}^{-2}} \mathrm{tr}\left(\varepsilon + \mathfrak{a}\mathfrak{b}^{-2}\mathfrak{b}'^{-2}, W(\underline{M}')^{\mathrm{new}}\right)$$
$$= \sum_{\mathfrak{b}''^2|\mathfrak{m}} \mathrm{tr}\left(\varepsilon + \mathfrak{a}\mathfrak{b}''^{-2}, W(\underline{M}')^{\mathrm{new}}\right) \sum_{\mathfrak{b}|\mathfrak{b}''} \mu(\mathfrak{b}).$$

Here $\underline{M}' = \underline{M}/\mathfrak{a}\mathfrak{b}^{-1}M/(\mathfrak{a}\mathfrak{b}^{-2}\mathfrak{b}'^{-1}(\mathfrak{a}\mathfrak{b}^{-1}M)^{\#}/\mathfrak{a}\mathfrak{b}^{-1}M)$. For the above identity we used the fact that the underlying module of $\underline{M}'$ is isomorphic to $\mathcal{O}/\mathfrak{a}\mathfrak{b}''^{-2}$ (which follows from Theorem 1.1 (ii) and Theorem 1.2 (ii)).

But the inner sum in the second identity equals zero unless $\mathfrak{b}'' = \mathcal{O}$. Therefore, the above identity equals $\mathrm{tr}\left(\varepsilon + \mathfrak{a}, W(\underline{M})^{\mathrm{new}}\right)$, i.e. (2.24) holds true. □

Proof of Theorem 2.5 (iii) By Proposition 2.19, the dimension of the $\mathrm{E}(\underline{M})$-module $W(\underline{M})^{\mathrm{new},\mathfrak{f}}$ is given by

$$\dim W(\underline{M})^{\mathrm{new},\mathfrak{f}} = \frac{1}{|\mathrm{E}(\underline{M})|} \sum_{g \in \mathrm{E}(\underline{M})} \psi_{\mathfrak{f}}(g) \chi_{W(\underline{M})^{\mathrm{new}}}(g),$$

where we used that $\psi_{\mathfrak{f}}(g)$, which is explained in Proposition 1.23, is real. We write the formula for $\chi_{W(\underline{M})^{\mathrm{new}}}(g)$ from Lemma 2.41 in the form

$$\chi_{W(\underline{M})^{\mathrm{new}}}(\varepsilon + \mathfrak{a}) = \prod_{\mathfrak{p}|\mathfrak{m}} I(\varepsilon, \mathfrak{p}),$$

$$I(\varepsilon, \mathfrak{p}) = \begin{cases} \mathrm{N}(\varepsilon - 1, \mathfrak{p}^a) & \text{if } \mathfrak{p}\|\mathfrak{m} \\ \mathrm{N}(\varepsilon - 1, \mathfrak{p}^a) - \mathrm{N}(\varepsilon - 1, \mathfrak{p}^{a-2}) & \text{if } \mathfrak{p}^2|\mathfrak{m}, \end{cases}$$

where $\mathfrak{p}^a$ is the exact power of $\mathfrak{p}$ dividing $\mathfrak{a}$. Inserting this quantity into the dimension formula we obtain

$$\dim W(\underline{M})^{\mathrm{new},\mathfrak{f}} = \frac{1}{|\mathrm{E}(\underline{M})|} \sum_{\varepsilon+\mathfrak{a} \in \mathrm{E}(\underline{M})} \psi_{\mathfrak{f}}(\varepsilon + \mathfrak{a}) \prod_{\mathfrak{p}|\mathfrak{m}} I(\varepsilon, \mathfrak{p}).$$

Using the decomposition of $\mathrm{E}(\underline{M})$ into $\mathfrak{p}$-parts as given in Proposition 1.22 we can write

$$\dim W(\underline{M})^{\mathrm{new},\mathfrak{f}} = \prod_{\mathfrak{p}|\mathfrak{m}} \frac{1}{2} \sum_{\varepsilon+\mathfrak{a}\in\langle\varepsilon_{\mathfrak{p}}+\mathfrak{a}\rangle} \psi_{\mathfrak{f}}(\varepsilon+\mathfrak{a})\, I(\varepsilon,\mathfrak{p}),$$

where $\varepsilon_{\mathfrak{p}} \equiv -1 \bmod \mathfrak{p}^a$, $\varepsilon_{\mathfrak{p}} \equiv +1 \bmod \mathfrak{a}\mathfrak{p}^{-a}$. (We used here that $I(\varepsilon,\mathfrak{p})$ depends only on ε modulo $\mathfrak{p}^a$, and that the order of $\mathrm{E}(\underline{M})$ equals 2^r, where r is the number of different prime factors of $\mathfrak{m}$.) We denote the factor corresponding to $\mathfrak{p}$ by $S(\mathfrak{p})$. Recall that $\psi_{\mathfrak{f}}(\varepsilon_{\mathfrak{p}}+\mathfrak{a}) = -1$ if $\mathfrak{p}|\mathfrak{f}$ and $\psi_{\mathfrak{f}}(\varepsilon_{\mathfrak{p}}+\mathfrak{a}) = +1$ otherwise (see Proposition 1.23 and the remark afterwards). In other words, $\psi_{\mathfrak{f}}(\varepsilon_{\mathfrak{p}}+\mathfrak{a}) = \mu(\mathfrak{f},\mathfrak{p})$. Inserting this and the formulas for $I(\varepsilon,\mathfrak{p})$ into the sum $S(\mathfrak{p})$ we obtain

$$S(\mathfrak{p}) = \frac{1}{2}\begin{cases} \mathrm{N}(\mathfrak{p}^a) + \mu(\mathfrak{f},\mathfrak{p})\,\mathrm{N}(2,\mathfrak{p}^a) & \text{if } \mathfrak{p}\|\mathfrak{m} \\ \mathrm{N}(\mathfrak{p}^a) - \mathrm{N}(\mathfrak{p}^{a-2}) + \mu(\mathfrak{f},\mathfrak{p})\big(\mathrm{N}(2,\mathfrak{p}^a) - \mathrm{N}(2,\mathfrak{p}^{a-2})\big) & \text{if } \mathfrak{p}^2|\mathfrak{m}. \end{cases}$$

It remains to prove that $\mathrm{N}(2,\mathfrak{p}^a) = \mathrm{N}(\mathfrak{p}^{a-1})$ if $\mathfrak{p}\|\mathfrak{m}$, and $\mathrm{N}(2,\mathfrak{p}^a) = \mathrm{N}(2,\mathfrak{p}^{a-2})$ if $\mathfrak{p}^2|\mathfrak{m}$. This is obvious if $\mathfrak{p}$ is odd.

If $\mathfrak{p}$ is even and $\mathfrak{p}\|\mathfrak{m}$, then $\mathfrak{p}^a\|\mathfrak{a} = \mathfrak{m}(2,\mathfrak{l})$ implies $\mathfrak{p}^{a-1}\|(2,\mathfrak{l})$; but by Proposition 1.13 we have $v_{\mathfrak{p}}(2,\mathfrak{l}) = v_{\mathfrak{p}}(2)$. Similarly, if $\mathfrak{p}$ is even and $\mathfrak{p}^2|\mathfrak{m}$, then $\mathfrak{p}^a\|\mathfrak{a} = \mathfrak{m}(2,\mathfrak{l})$ implies that $a-2 \geq v_{\mathfrak{p}}(2,\mathfrak{l})$, hence $a-2 \geq v_{\mathfrak{p}}(2)$. This proves the claimed formula. □

For the proof of the remaining part (ii) we need a lemma.

Lemma 2.42 *Let* $\mathfrak{m}$ *be an integral* $\mathcal{O}$*-ideal. The number of pairs* $(\mathfrak{b},\mathfrak{f})$ *of integral* $\mathcal{O}$*-ideals such that* $\mathfrak{b}^2|\mathfrak{m}$ *and* f *is a square-free divisor of* $\mathfrak{m}\mathfrak{b}^{-2}$ *equals* $\sigma_0(\mathfrak{m})$*, i.e. the number of integral* $\mathcal{O}$*-ideal divisors of* $\mathfrak{m}$.

Proof Denote the number of pairs $(\mathfrak{b},\mathfrak{f})$ in question by $I(\mathfrak{m})$. It is easy to see that the function I from the set of integral $\mathcal{O}$-ideals into $\mathbb{N}$ is multiplicative, i.e. it satisfies $I(\mathfrak{g}\mathfrak{h}) = I(\mathfrak{g})I(\mathfrak{h})$ for $\mathcal{O}$-ideals $\mathfrak{g}$ and $\mathfrak{h}$ with $(\mathfrak{g},\mathfrak{h}) = 1$. Hence, we have that $I(\mathfrak{m})$ equals $\prod_{\mathfrak{p}^n\|\mathfrak{m}} I(\mathfrak{p}^n)$. Since $\sigma_0(\mathfrak{m})$ is also multiplicative, it suffices to show that, for each prime ideal power $\mathfrak{p}^n$, we have $I(\mathfrak{p}^n) = \sigma_0(\mathfrak{p}^n)$. Indeed, $I(\mathfrak{p}^n)$ equals the number of pairs $(\mathfrak{p}^k,\mathcal{O})$ with $0 \leq 2k \leq n$ plus the number of pairs $(\mathfrak{p}^k,\mathfrak{p})$ with $0 \leq 2k < n$. Hence

$$I(\mathfrak{p}^n) = \begin{cases} 2(1+\lfloor\frac{n}{2}\rfloor) & \text{if } n \text{ is odd} \\ 1+2\lfloor\frac{n}{2}\rfloor & \text{if } n \text{ is even.} \end{cases}$$

We observe that in any case $I(\mathfrak{p}^n) = n+1$, for each $\mathfrak{p}$, which equals $\sigma_0(\mathfrak{p}^n)$. This proves the lemma. □

Proof of Theorem 2.5 (ii) The decomposition (2.21) given in (ii) is the decomposition of Theorem 2.4, which we specialize here to cyclic finite quadratic $\mathcal{O}$-modules.

Note that by the remark after Definition 2.40, the components in (ii) coincide with the ones given in Theorem 2.4.

It remains to prove that the components in (2.21) are irreducible. We shall prove in the next section (see Corollary 2.47) that the number of irreducible $\tilde{\Gamma}$-submodules of $W(\underline{M})$ is less than or equal to the number $\sigma_0(\mathfrak{m})$. If we insert the decompositions (2.21) for $\underline{M}/\mathfrak{a}\mathfrak{b}^{-1}M$ with $\mathfrak{b}$ running through the square divisors of $\mathfrak{m}$ into the decomposition (2.20), we have split $W(\underline{M})$ into as many $\tilde{\Gamma}$-submodules as there are pairs of integral $\mathcal{O}$-ideals $(\mathfrak{b}, \mathfrak{f})$ with $\mathfrak{b}^2|\mathfrak{m}$ and $\mathfrak{f}$ a square-free divisor of $\mathfrak{m}\mathfrak{b}^{-2}$. By Lemma 2.42 these are exactly $\sigma_0(\mathfrak{m})$-many $\tilde{\Gamma}$-submodules in the decomposition of $W(\underline{M})$, i.e. as many as our upper bound for irreducible submodules in $W(\underline{M})$. From the dimension formulas (2.22) it is clear that none of the components in this splitting $W(\underline{M})$ can be zero. Therefore the $\tilde{\Gamma}$-submodules in this splitting cannot split further and must hence be irreducible. This proves the theorem. □

2.5 The One Dimensional Subrepresentations

In the present section, we shall prove that, for a cyclic $\underline{M}$, the space $W(\underline{M})$ contains one-dimensional $\tilde{\Gamma}$-submodules if and only if the level of $\underline{M}$ is a character ideal (see the subsequent definition) times a square dividing the modified level of $\underline{M}$. Moreover, we shall also determine basis elements for the one-dimensional submodules of cyclic Weil representations.

Recall that if $\underline{M}$ is an $\mathcal{O}$-FQM with level $\mathfrak{l}$, the modified level of $\underline{M}$ equals $\mathfrak{l}(2, \mathfrak{l})^{-2}$, and the annihilator of $\underline{M}$ equals $\mathfrak{l}(2, \mathfrak{l})^{-1}$.

Definition 2.43 A *character ideal* is an integral $\mathcal{O}$-ideal $\mathfrak{c}$, which is of the form $\mathfrak{c} = \prod_{i=1}^{s} \mathfrak{p}_i \prod_{j=1}^{t} \mathfrak{q}_j^3$, where the $\mathfrak{p}_i$ are pairwise different prime ideals of degree one dividing 3, and where the $\mathfrak{q}_j$ are pairwise different prime ideals of degree and ramification index one dividing 2.

Remark Note that s or t might be equal to zero. If $t = 0$, then $\mathfrak{c}$ is called an *odd character ideal.*

Definition 2.44 For a prime ideal $\mathfrak{p}$ of degree one over 3, we use $\chi_{\mathfrak{p}}$ for the nontrivial Dirichlet character modulo $\mathfrak{p}$. For a prime ideal $\mathfrak{q}$ of degree one over 2, we use $\chi_{\mathfrak{q}^2}$ for the nontrivial Dirichlet character modulo $\mathfrak{q}^2$. For square-free products $\mathfrak{g}$ and $\mathfrak{h}$ of prime ideals of degree one over 3 and 2, respectively, we set

$$\chi_{\mathfrak{g}\mathfrak{h}^2} := \prod_{\mathfrak{p}|\mathfrak{g}} \chi_{\mathfrak{p}} \prod_{\mathfrak{q}|\mathfrak{h}} \chi_{\mathfrak{q}^2}, \tag{2.25}$$

and call $\chi_{\mathfrak{g}\mathfrak{h}^2}$ the *totally odd character modulo* $\mathfrak{g}\mathfrak{h}^2$.

Remark Note that, for primes $\mathfrak{p}$ and $\mathfrak{q}$ as in the definition, the groups of units $(\mathcal{O}/\mathfrak{p})^*$ and $(\mathcal{O}/\mathfrak{q}^2)^*$ have both order 2, so that there is indeed for each group a unique nontrivial character.

We state the main result of this section.

Theorem 2.6 *Let $\underline{M} = (M, Q)$ be an $\mathcal{O}$-CM with level $\mathfrak{l}$, annihilator $\mathfrak{a}$ and modified level $\mathfrak{m}$.*

(i) The space $W(\underline{M})$ contains one-dimensional $\tilde{\Gamma}$-submodules if and only if $\mathfrak{l}$ is a character ideal times a square dividing the modified level of $\underline{M}$.
(ii) The space $W(\underline{M})$ contains at most one one-dimensional $\tilde{\Gamma}$-submodule.
(iii) Suppose that $W(\underline{M})$ contains a one-dimensional $\tilde{\Gamma}$-submodule. If we write $\mathfrak{l} = \mathfrak{g}\mathfrak{h}^3\mathfrak{b}^2$, where $\mathfrak{g}\mathfrak{h}^3$ is the character ideal dividing $\mathfrak{l}$ and $\mathfrak{b}^2$ a divisor of the modified level of $\underline{M}$, then we have $\mathfrak{a} = \mathfrak{g}\mathfrak{h}^2\mathfrak{b}^2$ and $\mathfrak{m} = \mathfrak{g}\mathfrak{h}\mathfrak{b}^2$. The one-dimensional $\tilde{\Gamma}$-submodule equals $\iota_U W(\underline{M}/U)^{new,\mathfrak{g}\mathfrak{h}}$, where $U = \mathfrak{a}\mathfrak{b}^{-1}M = \mathfrak{g}\mathfrak{h}^2\mathfrak{b}M$. It is spanned by

$$\iota_U \sum_{s \in \mathcal{O}/\mathfrak{g}\mathfrak{h}^2} \chi_{\mathfrak{g}\mathfrak{h}^2}(s)\, e_{gs} = \sum_{\substack{x \in M, s \in \mathcal{O}/\mathfrak{g}\mathfrak{h}^2 \\ x \equiv s\gamma \bmod U}} \chi_{\mathfrak{g}\mathfrak{h}^2}(s)\, e_x. \tag{2.26}$$

Here $g = \gamma + U$ is a generator of $\underline{M}/U$, and $\chi_{\mathfrak{g}\mathfrak{h}^2}$ denotes the totally odd Dirichlet character modulo $\mathfrak{g}\mathfrak{h}^2$.

The rest of this section is devoted to the proof of the theorem. For this it is convenient to introduce a name for the prime ideals occurring in character ideals.

Definition 2.45 A prime ideal of degree 1 and ramification index 1 above 2 is called a $(2, 1, 1)$-*ideal*. A prime ideal of degree 1 above 3 is called a $(3, 1)$-*ideal*.

Thus, a character ideal $\mathfrak{c}$ is a product of different $(3, 1)$-ideals and cubes of different $(2, 1, 1)$-ideals. For proving the theorem we first consider the new parts of the spaces $W(\underline{M})$.

Lemma 2.46 *Let $\underline{M}$ be an $\mathcal{O}$-CM with level $\mathfrak{l}$. The space $W(\underline{M})^{new}$ contains one-dimensional $\tilde{\Gamma}$-submodules if and only if $\mathfrak{l}$ is a character ideal. If $\mathfrak{l}$ is a character ideal, say, $\mathfrak{l} = \mathfrak{g}\mathfrak{h}^3$, then $W(\underline{M})^{new}$ contains exactly one one-dimensional $\tilde{\Gamma}$-submodule, namely $W(\underline{M})^{new,\mathfrak{g}\mathfrak{h}}$.*

Proof First suppose that $\mathfrak{l}$ is a character ideal such that $\mathfrak{l} = \mathfrak{g}\mathfrak{h}^3$. Then the modified level of $\underline{M}$ equals $\mathfrak{g}\mathfrak{h}$. Hence, by Theorem 2.5 (iii) we have that $W(\underline{M})^{\mathrm{new},\mathfrak{g}\mathfrak{h}}$ is one dimensional, i.e. $W(\underline{M})$ contains one-dimensional $\tilde{\Gamma}$-submodules.

Next suppose that $W(\underline{M})^{\mathrm{new}}$ contains one-dimensional $\tilde{\Gamma}$-submodules. Let $\mathfrak{m}$ be the modified level of $\underline{M}$. By Theorem 2.5 (ii) (and by Proposition 2.16) there exists a square-free divisor $\mathfrak{f}$ of $\mathfrak{m}$ such that $W(\underline{M})^{\mathrm{new},\mathfrak{f}}$ is one dimensional. If we can show that $\mathfrak{l} = \mathfrak{g}\mathfrak{h}^3$ and $\mathfrak{f} = \mathfrak{g}\mathfrak{h}$ for a product $\mathfrak{g}$ of different $(3, 1)$-ideals and a product $\mathfrak{h}$ of different $(2, 1, 1)$-ideals, then this proves the lemma.

We write $\dim W(\underline{M})^{\text{new},\mathfrak{f}} = P_1 P_2$, where P_1 and P_2 are the contributions from odd and even prime ideals, respectively. By the assumption that $W(\underline{M})^{\text{new},\mathfrak{f}}$ is one dimensional, we have $P_1 = 1$ and $P_2 = 1$ (see also Proposition 1.6). Using (2.22) and also the fact that $\mathfrak{a} = \mathfrak{m}(2,\mathfrak{l})$, we can write

$$1 = P_1 = \mathrm{N}(\mathfrak{m}_1) \prod_{\mathfrak{p}\|\mathfrak{m}_1} \mathrm{N}(\mathfrak{p})^{-1} \prod_{\mathfrak{p}\|\mathfrak{m}_1} \left(\frac{\mathrm{N}(\mathfrak{p}) + \mu(\mathfrak{f},\mathfrak{p})}{2}\right) \times \prod_{\mathfrak{p}^2|\mathfrak{m}_1} \mathrm{N}(\mathfrak{p})^{-2} \prod_{\mathfrak{p}^2|\mathfrak{m}_1} \left(\frac{\mathrm{N}(\mathfrak{p})^2 - 1}{2}\right), \tag{2.27}$$

where $\mathfrak{m}_1$ stands for the odd part of $\mathfrak{m}$. Since the second and the forth products and also $\mathrm{N}(\mathfrak{m}_1)$ times the first and the third products in (2.27) are all integers, obviously we need to have first of all that $\mathfrak{m}$ is square-free. Moreover, for all $\mathfrak{p}\|\mathfrak{m}_1$, we need to have

$$\frac{\mathrm{N}(\mathfrak{p}) + \mu(\mathfrak{f},\mathfrak{p})}{2} = 1.$$

But this implies that $\mathrm{N}(\mathfrak{p}) = 3$ and $\mu(\mathfrak{f},\mathfrak{p}) = -1$ for each $\mathfrak{p}\|\mathfrak{m}_1$. Therefore, we have that $\mathfrak{m}_1 = \mathfrak{g}$, and the odd part of $\mathfrak{f}$ equals $\mathfrak{g}$, where $\mathfrak{g}$ is a product of different $(3,1)$-ideals.

Now we consider the even part. Using (2.22), we have

$$1 = P_2 = 2^{-s}\,\mathrm{N}\left(\mathfrak{m}_2(2,\mathfrak{l})\right) \prod_{\mathfrak{p}\|\mathfrak{m}_2} \mathrm{N}(\mathfrak{p})^{-1} \prod_{\mathfrak{p}\|\mathfrak{m}_2} \left(\mathrm{N}(\mathfrak{p}) + \mu(\mathfrak{f},\mathfrak{p})\right) \times \prod_{\mathfrak{p}^2|\mathfrak{m}_2} \mathrm{N}(\mathfrak{p})^{-2} \prod_{\mathfrak{p}^2|\mathfrak{m}_2} \left(\mathrm{N}(\mathfrak{p})^2 - 1\right), \tag{2.28}$$

where $\mathfrak{m}_2$ denotes the even part of $\mathfrak{m}$, and s denotes the number of distinct prime ideal divisors of $\mathfrak{m}_2$. First note that the second and the forth products in (2.28) are integers. Also $2^{-s}\,\mathrm{N}(\mathfrak{m}_2(2,\mathfrak{l}))$ times the first and the third products in (2.28) are integers. Indeed, this follows from the fact that every prime ideal dividing $\mathfrak{m}_2$ occurs in the prime ideal decomposition of $(2,\mathfrak{l})$, since $\mathfrak{m} = \mathfrak{l}(2,\mathfrak{l})^{-2}$. Therefore, we need to have first of all that $\mathfrak{m}_2$ is square-free. Furthermore, we need to have

$$\mathrm{N}(2,\mathfrak{l}) = 2^s, \quad \mathrm{N}(\mathfrak{p}) + \mu(\mathfrak{f},\mathfrak{p}) = 1$$

for all $\mathfrak{p}\|\mathfrak{m}_2$. But the first identity implies that $\mathfrak{m}_2 = (2,\mathfrak{l}) = \mathfrak{h}$, where $\mathfrak{h}$ is a product of different $(2,1,1)$-ideals. The second identity implies that $\mathrm{N}(\mathfrak{p}) = 2$ and $\mu(\mathfrak{f},\mathfrak{p}) = -1$ for each $\mathfrak{p}\|\mathfrak{m}_2$. Therefore, we have that $\mathfrak{m}_2 = \mathfrak{h}$ and that the even part of $\mathfrak{f}$ equals $\mathfrak{h}$.

As a consequence, we obtain $\mathfrak{m} = \mathfrak{m}_1\mathfrak{m}_2 = \mathfrak{g}\mathfrak{h}$, $\mathfrak{f} = \mathfrak{g}\mathfrak{h}$, and hence we have that $\mathfrak{l} = \mathfrak{m}(2,\mathfrak{l})^2 = \mathfrak{g}\mathfrak{h}\mathfrak{h}^2 = \mathfrak{g}\mathfrak{h}^3$, which proves the lemma. □

Remark Note that if $\underline{M}$ is anisotropic, i.e. $\underline{M}$ does not contain isotropic submodules, then $W(\underline{M})$, which equals $W(\underline{M})^{\text{new}}$, contains one-dimensional $\tilde{\Gamma}$-submodules if and only if the level of $\underline{M}$ is a character ideal (see Lemma 2.46).

Proof of Theorem 2.6

Proof of part (i). Suppose that the space $W(\underline{M})$ contains one-dimensional $\tilde{\Gamma}$-submodules. By Theorem 2.5 (and Proposition 2.16), there exists an integral $\mathcal{O}$-ideal $\mathfrak{b}$ with $\mathfrak{b}^2|\mathfrak{m}$ such that the space $W(\underline{M}/\mathfrak{a}\mathfrak{b}^{-1}M)^{\text{new},\mathfrak{f}}$ is one dimensional. Lemma 2.46 implies that the level of $\underline{M}/\mathfrak{a}\mathfrak{b}^{-1}M$ is a character ideal. But the level of $\underline{M}/\mathfrak{a}\mathfrak{b}^{-1}M$ equals $\mathfrak{l}\mathfrak{b}^{-2}$ (Corollary 1.19). Hence $\mathfrak{l}$ is of the claimed form.

Suppose that $\mathfrak{l}$ is as given in the statement of the theorem, i.e $\mathfrak{l} = \mathfrak{c}\mathfrak{b}^2$, where $\mathfrak{b}^2|\mathfrak{m}$ and $\mathfrak{c}$ is a character ideal. Set $U := \mathfrak{a}\mathfrak{b}^{-1}M$. Since the level of $\underline{M}/U$ equals $\mathfrak{l}\mathfrak{b}^{-2} = \mathfrak{c}$ (see Corollary 1.19), which is a character ideal, we deduce from Lemma 2.46 that the space $W(\underline{M}/U)^{\text{new}}$, and hence the space $\iota_U W(\underline{M}/U)^{\text{new}} \subseteq W(\underline{M})$ contains one-dimensional $\tilde{\Gamma}$-submodules.

Proof of part (ii). First we show that amongst the $\tilde{\Gamma}$-submodules in the decomposition of $W(\underline{M})$ obtained on combining (2.20) and (2.21), there is at most one one-dimensional $\tilde{\Gamma}$-submodule. Suppose $W(\underline{M})$ contains two one-dimensional $\tilde{\Gamma}$-submodules in the decomposition, say $W(\underline{M}/\mathfrak{a}\mathfrak{b}_i^{-1}M)^{\text{new},\mathfrak{f}_i}$ $(i = 1, 2)$. Then, by Lemma 2.46, the level $\mathfrak{l}_i$ of $\underline{M}/\mathfrak{a}\mathfrak{b}_i^{-1}M$ is equal to a character ideal, say $\mathfrak{g}_i\mathfrak{h}_i^3$, and $\mathfrak{f}_i = \mathfrak{g}_i\mathfrak{h}_i$. From Corollary 1.19, we know that $\mathfrak{l}_i = \mathfrak{l}\mathfrak{b}_i^{-2}$. Hence, $\mathfrak{l} = \mathfrak{g}_i\mathfrak{h}_i^3\mathfrak{b}_i^2$. But this implies that $\mathfrak{g}_1 = \mathfrak{g}_2$, $\mathfrak{h}_1 = \mathfrak{h}_2$ and $\mathfrak{b}_1 = \mathfrak{b}_2$ (use that $\mathfrak{g}_1\mathfrak{h}_1 = \mathfrak{g}_2\mathfrak{h}_2$ is the square-free part of the unique factorization of $\mathfrak{l}$ into a product of a square-free ideal and a square), i.e. the claimed result holds true.

Now suppose that W is a one-dimensional $\tilde{\Gamma}$-submodule of $W(\underline{M})$. Then by Proposition 2.16, we have $W \simeq W(\underline{M}/\mathfrak{a}\mathfrak{b}^{-1}M)^{\text{new},\mathfrak{f}}$, for some $\mathfrak{b}$ and $\mathfrak{f}$ as in Eqs. (2.20) and (2.21). If $W(\underline{M})$ does not contain any one-dimensional $\tilde{\Gamma}$-submodule in the decomposition, there is nothing to prove. If there is a one-dimensional $\tilde{\Gamma}$-submodule in the decomposition of $W(\underline{M})$, it is unique by the above argument. We call it $W(\underline{M}/\mathfrak{a}\mathfrak{b}_1^{-1}M)^{\text{new},\mathfrak{f}_1}$. Hence, all the other $\tilde{\Gamma}$-submodules $W(\underline{M}/\mathfrak{a}\mathfrak{b}^{-1}M)^{\text{new},\mathfrak{f}}$ with $\mathfrak{b} \neq \mathfrak{b}_1$ or $\mathfrak{f} \neq \mathfrak{f}_1$ have dimension bigger than one. Our aim is to show that $W = W(\underline{M}/\mathfrak{a}\mathfrak{b}_1^{-1}M)^{\text{new},\mathfrak{f}_1}$. We denote by $P_{\mathfrak{b},\mathfrak{f}}$ the projection from W onto the space $W(\underline{M}/\mathfrak{a}\mathfrak{b}^{-1})^{\text{new},\mathfrak{f}}$. It suffices to prove for all $w \in W$, the identity $P_{\mathfrak{b}_1,\mathfrak{f}_1}(w) = w$. The identity $\sum_{\mathfrak{b},\mathfrak{f}} P_{\mathfrak{b},\mathfrak{f}} = 1$ implies that $w = \sum_{\mathfrak{b},\mathfrak{f}} P_{\mathfrak{b},\mathfrak{f}}(w)$. But $P_{\mathfrak{b},\mathfrak{f}}(w) = 0$ for all $(\mathfrak{b},\mathfrak{f}) \neq (\mathfrak{b}_1,\mathfrak{f}_1)$. Indeed, the kernel of the map $P_{\mathfrak{b},\mathfrak{f}}|_W$ must be equal to W, since otherwise the map $P_{\mathfrak{b},\mathfrak{f}}|_W$ would be a $\tilde{\Gamma}$-linear isomorphism from W onto a one-dimensional $\tilde{\Gamma}$-submodule of $W(\underline{M}/\mathfrak{a}\mathfrak{b}^{-1}M)^{\text{new},\mathfrak{f}}$, whereas the latter is irreducible and has dimension bigger than one. Hence, we have $w = P_{\mathfrak{b}_1,\mathfrak{f}_1}(w)$, which proves (ii).

Proof of part (iii). Suppose $W(\underline{M})$ contains a one-dimensional $\tilde{\Gamma}$-submodule, say W. As we saw in the proof of part (ii), we then have $\mathfrak{l} = \mathfrak{g}\mathfrak{h}^3\mathfrak{b}^2$ with $\mathfrak{b}^2|\mathfrak{m}$, and $W = \iota_U W(\underline{M}/U)^{\text{new},\mathfrak{g}\mathfrak{h}}$, where $U = \mathfrak{a}\mathfrak{b}^{-1}M$.

For proving the claimed identities for $\mathfrak{a}$ and $\mathfrak{m}$ it suffices to show that $\mathfrak{h} = (2, \mathfrak{l})$ (since, for any $\mathcal{O}$-CM, we have $\mathfrak{a} = \mathfrak{l}(2, \mathfrak{l})^{-1}$ and $\mathfrak{m} = \mathfrak{l}(2, \mathfrak{l})^{-2}$). For this we write $\mathfrak{m} = \mathfrak{b}^2\mathfrak{t}$. Since $\mathfrak{m} = \mathfrak{l}(2, \mathfrak{l})^{-2}$ we have $(2, \mathfrak{l})^2 = \mathfrak{g}\mathfrak{h}^3\mathfrak{t}^{-1}$, and since $\mathfrak{g}$ and $\mathfrak{h}$ are square-free and relatively prime, therefore $(2, \mathfrak{l})|\mathfrak{h}$. But $\mathfrak{h}$ divides 2 and it divides $\mathfrak{l}$, hence $(2, \mathfrak{l}) = \mathfrak{h}$.

Finally, let $I := \sum_{s \in \mathcal{O}/\mathfrak{g}\mathfrak{h}^2} \chi_{\mathfrak{g}\mathfrak{h}^2}(s)\, e_{gs}$, where $g = \gamma + U$ is a generator of the $\mathcal{O}$-CM $\underline{M}' = \underline{M}/U \simeq \mathcal{O}/\mathfrak{g}\mathfrak{h}^2$. Since I is clearly different from 0, it remains to show that I is in $W(\underline{M}')^{\text{new},\mathfrak{g}\mathfrak{h}}$. First of all, note that $W(\underline{M}')^{\text{new}} = W(\underline{M}')$ since $\underline{M}'$ has modified level $\mathfrak{g}\mathfrak{h}$, and hence has no isotropic submodules different from zero (see Theorem 1.2). In other words, we only have to show that $hI = \psi_{\mathfrak{g}\mathfrak{h}}(h)I$ for all h in $E(\underline{M}')$. But this follows immediately from the fact that $E(\underline{M}') = (\mathcal{O}/\mathfrak{g}\mathfrak{h}^2)^*$ and $\psi_{\mathfrak{g}\mathfrak{h}} = \chi_{\mathfrak{g}\mathfrak{h}^2}$. This proves the theorem. □

2.6 The Number of Irreducible Components

In the present section we shall find an estimate for the number of irreducible subrepresentations of Weil representations. For cyclic Weil representations this number can be made even more explicit. Namely, we have:

Theorem 2.7 *Let $\underline{M} = (M, Q)$ be an $\mathcal{O}$-FQM with level $\mathfrak{l}$. Then the number of irreducible $\tilde{\Gamma}$-submodules of $W(\underline{M})$ is less than or equal to the number of elements of $(M \times M)'/\Gamma_{\mathcal{O}/\mathfrak{l}}$. Here $(M \times M)'$ is the set of all v in $M \times M$ such that χ_v is trivial (for χ_v, we refer to Lemma 2.58 below).*

Corollary 2.47 *Let $\underline{M} = (M, Q)$ be an $\mathcal{O}$-CM with modified level $\mathfrak{m}$. The number of irreducible $\tilde{\Gamma}$-submodules of $W(\underline{M})$ is less than or equal to the number of integral $\mathcal{O}$-ideal divisors of $\mathfrak{m}$, i.e. $\sigma_0(\mathfrak{m})$.*

We prove the above theorem with two different methods. This section is divided accordingly into two subsections. Both approaches calculate the dimensions of the intertwining algebras of Weil representations. In fact, the number of irreducible G-submodules of a G-module V is bounded by the dimension of the intertwining algebra of V (see Proposition 2.22).

As a side result of our second approach we also obtain the following theorem:

Theorem 2.8 *Let $\underline{M} = (M, Q)$ be an $\mathcal{O}$-FQM with level $\mathfrak{l}$ and associated bilinear form B. There exists a projective representation ρ of Γ which satisfies, for any $z \in M$, the following formulas*

(i) $\rho(T_b)e_z = e\{bQ(z)\}\, e_z \quad (b \in \mathcal{O})$
(ii) $\rho(S)e_z = \sigma(\underline{M}) \frac{1}{\sqrt{|M|}} \sum_{z' \in M} e\{-B(z', z)\}\, e_{z'}$.

2.6.1 The First Approach

Before we can give the proofs of Theorem 2.7 and Corollary 2.47, we need several lemmas.

Lemma 2.48 *Let $\underline{M} = (M, Q)$ be an $\mathcal{O}$-FQM. The bilinear map*

$$[\_,\_] : W(\underline{M}^{-1}) \times W(\underline{M}) \to \mathbb{C}, \_\left[\sum_{x\in M} v(x)e_x, \sum_{x'\in M} v'(x')e_{x'}\right] := \sum_{x\in M} v(x)v'(x)$$

is $\tilde{\Gamma}$-invariant.

Proof Let B be the bilinear form of $\underline{M}$. It is enough to prove the lemma for the standard generators T_b^* ($b \in \mathcal{O}$) and S^*. We shall prove only the invariance under S^*, since the invariance under T_b^* is obvious. Write $v = \sum_{x\in M} v(x)e_x$ and $v' = \sum_{x'\in M} v'(x')e_{x'}$. From the S^*-action in (2.11), we have

$$S^*v = \frac{\sigma(\underline{M}^{-1})}{\sqrt{|M|}} \sum_{x\in M} v(x) \sum_{y\in M} e\{B(y, x)\}\, e_y$$

and

$$S^*v' = \frac{\sigma(\underline{M})}{\sqrt{|M|}} \sum_{x'\in M} v'(x) \sum_{y'\in M} e\{-B(y', x')\}\, e_{y'}.$$

Hence, we have

$$[S^*v, S^*v'] = \frac{\sigma(\underline{M}^{-1})\sigma(\underline{M})}{|M|} \sum_{x,x'\in M} v(x)v'(x') \sum_{y\in M} e\{B(y, x - x')\}.$$

From Proposition 1.11, the inner sum is zero unless $x = x'$, otherwise it equals $|M|$. We now recognize $[S^*v, S^*v'] = [v, v']$, since Proposition 1.10 implies that sigma-invariant $\sigma(\underline{M})$ has absolute value one. □

Lemma 2.49 *Let $\underline{M} = (M, Q)$ be an $\mathcal{O}$-FQM. Then the linear map*

$$W(\underline{M}^{-1}) \to W(\underline{M})^\bullet, \quad v \mapsto \text{“}v' \mapsto [v, v']\text{''}$$

defines a $\tilde{\Gamma}$-module isomorphism.

Proof We denote the above map by φ. First we show that φ is $\tilde{\Gamma}$-linear. For that, using (2.4), it is enough to show that for any $v \in W(\underline{M}^{-1})$, $v' \in W(\underline{M})^\bullet$ and α in $\tilde{\Gamma}$, the identity $[\alpha^{-1}v, v'] = [v, \alpha v']$ holds true. But if do the substitution $v \mapsto \alpha v$, Lemma 2.48 implies the result.

To show that φ is an isomorphism, it suffices to show that φ is an injection, since the spaces have the same dimension. Let v be an element of the kernel of φ, i.e. $[v, v'] = 0$ for all $v' \in W(\underline{M})^{\bullet}$. We write $v = \sum_{x \in M} v(x) e_x$ and $v' = \sum_{x' \in M} v'(x') e_{x'}$. Then we have

$$[v, v'] = \sum_{x \in M} v(x) v'(x) = 0.$$

If we take $v' = e_{x_0}$ for some $x_0 \in M$, then the above identity implies that $v(x_0) = 0$. Repeating the same argument by choosing other elements of $\underline{M}$, we observe that $v = 0$, i.e. φ is an injection. □

Lemma 2.50 *Let $\underline{M} = (M, Q)$, $\underline{N} = (N, R)$ be $\mathscr{O}$-FQM. Then the linear map*

$$W(\underline{M} + \underline{N}) \to W(\underline{M}) \otimes W(\underline{N}), \quad e_{x \oplus y} \mapsto e_x \otimes e_y$$

defines a $\tilde{\Gamma}$-module isomorphism.

Proof We denote the map in the lemma by φ. Clearly φ is an isomorphism of vector spaces $\mathbb{C}[M \oplus N]$ and $\mathbb{C}[M] \otimes \mathbb{C}[N]$. Hence it remains to show that φ is $\tilde{\Gamma}$-linear. It is enough to prove this fact for T_b^* ($b \in \mathscr{O}$) and S^*. Let QR denote the quadratic form on $\underline{M} + \underline{N}$. Recall that $QR(x \oplus y) = Q(x) + R(y)$. Let $b \in \mathscr{O}$, $x \in M$ and $y \in N$. The T_b^*-action in (2.11) and φ commute, since

$$\begin{aligned} \varphi(T_b^* e_{x \oplus y}) &= e\{bQR(x \oplus y)\} e_x \otimes e_y = e\{b(Q(x) + R(y))\} e_x \otimes e_y \\ &= e\{bQ(x)\} e_x \otimes e\{bR(y)\} e_y = T_b^* e_x \otimes T_b^* e_y = T_b^*(e_x \otimes e_y) \\ &= T_b^* \varphi(e_{x \oplus y}). \end{aligned}$$

Let B, C and BC stand for associated bilinear forms of $\underline{M}$, $\underline{N}$ and $\underline{M} + \underline{N}$, respectively. Recall $BC(x' \oplus y', x \oplus y) = B(x', x) + C(y', y)$. Similarly, the identity

$$\begin{aligned} \varphi(S^* e_{x \oplus y}) &= \frac{\sigma(\underline{M} + \underline{N})}{|\underline{M} + \underline{N}|} \sum_{x' \oplus y' \in \underline{M} \oplus \underline{N}} e\left\{-BC(x' \oplus y', x \oplus y)\right\} e_x \otimes e_y \\ &= \frac{\sigma(\underline{M} + \underline{N})}{|\underline{M} + \underline{N}|} \sum_{x' \in \underline{M}} e\left\{-B(x', x)\right\} e_x \otimes \sum_{y' \in N} e\left\{-C(y', y)\right\} e_y \\ &= S^* \varphi(e_{x \oplus y}) \end{aligned}$$

proves that the S^*-action in (2.11) and φ commute. For the last identity we used the remark after Definition 1.8, which states that $\sigma(\underline{M} + \underline{N}) = \sigma(\underline{M}) \sigma(\underline{N})$, and we also used $|\underline{M} + \underline{N}| = |M||N|$. □

Lemma 2.51 *Let $\underline{M} = (M, Q)$ be an $\mathcal{O}$-FQM. Then*

$$\mathrm{Hom}_{\tilde{\Gamma}}\left(W(\underline{M}), W(\underline{M})\right) \simeq W\left(\underline{M}^{-1} + \underline{M}\right)^{\tilde{\Gamma}}.$$

Proof From the map given in (2.5), it is easy to see that $W(\underline{M})^{\bullet} \otimes W(\underline{M})$ and $\mathrm{Hom}\left(W(\underline{M}), W(\underline{M})\right)$ are isomorphic as $\tilde{\Gamma}$-modules. Lemma 2.49 implies that the spaces $W(\underline{M})^{\bullet}$ and $W(\underline{M}^{-1})$ are $\tilde{\Gamma}$-module isomorphic. Using Lemma 2.50 we obtain that $W(\underline{M}^{-1}) \otimes W(\underline{M})$ is $\tilde{\Gamma}$-module isomorphic to $W\left(\underline{M}^{-1} + \underline{M}\right)$. Therefore, we have that $\mathrm{Hom}\left(W(\underline{M}), W(\underline{M})\right)$ is $\tilde{\Gamma}$-module isomorphic to $W\left(\underline{M}^{-1} + \underline{M}\right)$. Therefore Proposition 2.21 implies the result. □

Lemma 2.52 *Let $\underline{M} = (M, Q)$ be an $\mathcal{O}$-FQM with annihilator $\mathfrak{a}$. The group $\Gamma_{\mathcal{O}/\mathfrak{a}}$ acts on the right of $M \times M$ via*

$$\left((x, y), A = \left(\begin{smallmatrix} a+\mathfrak{a} & b+\mathfrak{a} \\ c+\mathfrak{a} & d+\mathfrak{a} \end{smallmatrix}\right)\right) \mapsto (x, y)A, \quad (x, y)A := (ax + cy, bx + dy).$$

Proof First we show that the above multiplication is well-defined. Let $a' \in a + \mathfrak{a}$. We have $a' = a + t$ for some $t \in \mathfrak{a}$. But $a'x = (a+t)x = ax$, since $tx = 0$. Let $v = (x, y) \in M \times M$ and $A, B \in \Gamma_{\mathcal{O}/\mathfrak{a}}$. Write $A = \left(\begin{smallmatrix} a+\mathfrak{a} & b+\mathfrak{a} \\ c+\mathfrak{a} & d+\mathfrak{a} \end{smallmatrix}\right)$ and $B = \left(\begin{smallmatrix} a'+\mathfrak{a} & b'+\mathfrak{a} \\ c'+\mathfrak{a} & d'+\mathfrak{a} \end{smallmatrix}\right)$. Since it is obvious that $v1 = v$, the following identity proves the lemma:

$$B(vA) = \left(a'ax + a'cy + c'bx + c'dy, b'ax + b'cy + d'bx + d'dy\right) = ABv.$$

□

Remark Let $\underline{M} = (M, Q)$ be an $\mathcal{O}$-FQM with level $\mathfrak{l}$ and annihilator $\mathfrak{a}$. By Proposition 1.5, we have that $\mathfrak{l} \subseteq \mathfrak{a}$, so there is a reduction map from $\mathcal{O}/\mathfrak{l}$ onto $\mathcal{O}/\mathfrak{a}$. Hence, using Proposition 2.2 and Lemma 2.52, we obtain that $\Gamma_{\mathcal{O}/\mathfrak{l}}$ also acts on $M \times M$.

Lemma 2.53 *Let $\underline{M} = (M, Q)$ be an $\mathcal{O}$-FQM with level $\mathfrak{l}$. For fixed $A = \left(\begin{smallmatrix} a+\mathfrak{l} & b+\mathfrak{l} \\ c+\mathfrak{l} & d+\mathfrak{l} \end{smallmatrix}\right)$ in $\Gamma_{\mathcal{O}/\mathfrak{l}}$, the map*

$$f_A : M \times M \to \mathbb{C}^*, \quad (x, y) \mapsto e\left\{-\left(abQ(x) + bcB(x, y) + cdQ(y)\right)\right\}$$

satisfies the following identity

$$f_{AB}(v) = f_A(v) f_B(vA).$$

Here $(v, A) \mapsto vA$ is the action in Lemma 2.52 (see also the remark afterwards).

Proof First note that $f_A(v)$ depends only on the coset of a. Let $a' \in a + \mathfrak{l}$. We have $a' = a + l$ for some $l \in \mathfrak{l}$. But $a'Q(x) = (a + l)Q(x) = aQ(x)$ for any $x \in \underline{M}$. Let $B = \begin{pmatrix} a'+\mathfrak{l} & b'+\mathfrak{l} \\ c'+\mathfrak{l} & d'+\mathfrak{l} \end{pmatrix}$ be in $\Gamma_{\mathcal{O}/\mathfrak{l}}$ and $v = (x, y)$ be in $M \times M$. Let B stand for associated bilinear form of $\underline{M}$. The following proves the claimed identity:

$$\begin{aligned} f_A(v) f_B(vA) &= e\left\{-\left(abQ(x) + bcB(x,y) + cdQ(y)\right)\right\} \times \\ &\quad \times e\left\{-\left(a'b'Q(ax+cy) + b'c'B(ax+cy, bx+dy) + c'd'Q(bx+dy)\right)\right\} \\ &= e\left\{-\left(ab + a^2ab' + 2abd'c' + c'd'b^2\right)Q(x)\right\} \times \\ &\quad \times e\left\{-\left(cd + a'b'c^2 + 2a'c'cd + c'd'd^2\right)Q(y)\right\} \times \\ &\quad \times e\left\{-\left(bc + a'b'ac + a'c'(ad+bc) + c'd'bd\right)B(x,y)\right\} \\ &= e\left\{-(ad' + bc')(ab' + bd')Q(x)\right\} \times \\ &\quad \times e\left\{-(ab' + bd')(ca' + dc')B(x,y)\right\} e\left\{-(ca' + dc')(cb' + dd')Q(y)\right\} \\ &= f_{AB}(v). \end{aligned}$$

For the third identity we used $a'd' - b'c' \equiv 1 \bmod \mathfrak{l}$ and that $ad - bc \equiv 1 \bmod \mathfrak{l}$, and, in addition, the fact that $lQ(x) = 0$ for any $x \in \underline{M}$, $l \in \mathfrak{l}$. □

Lemma 2.54 *Let $\underline{M} = (M, Q)$ be an $\mathcal{O}$-FQM with level $\mathfrak{l}$. The group $\Gamma_{\mathcal{O}/\mathfrak{l}}$ acts on $\mathbb{C}[M \times M]$ via*

$$(A, e_v) \mapsto Ae_v, \quad Ae_v := \overline{f_{A^{-1}}(v)}e_{vA^{-1}},$$

where f_A is as in Lemma 2.53.

Proof Let A, B be in $\Gamma_{\mathcal{O}/\mathfrak{l}}$ and v be in $M \times M$. Since it is obvious that $1e_v = e_v$, the following identity proves the lemma:

$$A(Be_v) = f_{B^{-1}}(v) f_{A^{-1}}((vB)^{-1}) e_{(vB)^{-1}A^{-1}} = f_{(AB)^{-1}}(v) e_{v(AB)^{-1}} = ABe_v.$$

For the second identity we used Lemma 2.53. □

Definition 2.55 Let $\underline{M} = (M, Q)$ be an $\mathcal{O}$-FQM level $\mathfrak{l}$. In the following the $\Gamma_{\mathcal{O}/\mathfrak{l}}$-module $\mathbb{C}[M \times M]$ described in Lemma 2.54 is denoted by $P(\underline{M})$.

Remark Let $\mathfrak{a}$ be a nonzero integral $\mathcal{O}$-ideal. There is an epimorphism from $\tilde{\Gamma}$ onto $\Gamma_{\mathcal{O}/\mathfrak{a}}$ which maps T_b^* ($b \in \mathcal{O}$) and S^* to T_b and S^* reduced modulo $\mathfrak{a}$, respectively. This epimorphism obtained by composing the epimorphism from $\tilde{\Gamma}$ onto Γ (see Sect. 2.2) and the epimorphism from Γ onto $\Gamma_{\mathcal{O}/\mathfrak{a}}$ (see Lemma 1.29).

Remark The above remark and Proposition 2.2 imply that $P(\underline{M})$ can be viewed as a $\tilde{\Gamma}$-module.

Lemma 2.56 *Let* $\underline{M} = (M, Q)$ *be an* $\mathcal{O}$*-FQM with bilinear form* B *and level* $\mathfrak{l}$. *The linear map*

$$W(\underline{M}^{-1} + \underline{M}) \to P(\underline{M}), \quad \kappa : e_{x\oplus y} \mapsto \sum_{z\in M} e\{B(z,y)\}\, e_{(y-x,z)}$$

defines a $\tilde{\Gamma}$*-module isomorphism.*

Proof First note by the second remark after Definition 2.55 that $P(\underline{M})$ is a $\tilde{\Gamma}$-module. It is clear that the map κ is an isomorphism. It remains to show that κ is $\tilde{\Gamma}$-linear, i.e. $\kappa\alpha e_{x\oplus y} = \pi(\alpha)\kappa e_{x\oplus y}$ for $\alpha = T_b^*$ ($b \in \mathcal{O}$) or S^*. Here π stands for the epimorphism from $\tilde{\Gamma}$ onto $\Gamma_{\mathcal{O}/\mathfrak{l}}$ explained in the first remark after Definition 2.55. Let $b \in \mathcal{O}$. For T_b^* the claimed identity holds true, since for any $x, y \in M$, we have

$$\begin{aligned}
\pi(T_b^*)\kappa e_{x\oplus y} &= \sum_{z\in M} e\{B(z,y)\}\, \overline{f_{\pi(T_b^*)^{-1}}(y-x,z)}e_{(y-x,z)\pi(T_b^*)^{-1}}\\
&= e\{-bQ(y-x)\} \sum_{z\in M} e\{B(z,y)\}\, e_{(y-x,b(x-y)+z)}\\
&= e\{-bQ(y-x) + bB(y,y-x)\} \sum_{z\in M} e\{B(z,y)\}\, e_{(y-x,z)}\\
&= e\big\{b\big(Q(y) - Q(x)\big)\big\} \sum_{z\in M} e\{B(z,y)\}\, e_{(y-x,z)} = \kappa T_b^* e_{x\oplus y}.
\end{aligned}$$

To obtain the third identity we did the substitution $z \mapsto z - b(y-x)$ in the previous sum. We refer to the T_b^*-action in (2.11) to see that the last identity holds true.

We have

$$\begin{aligned}
\pi(S^*)\kappa e_{x\oplus y} &= \sum_{z\in M} e\{B(z,y)\}\, \overline{f_{\pi(S^*)^{-1}}(y-x,z)}e_{(y-x,z)\pi(S^*)^{-1}}\\
&= \sum_{z\in M} e\{B(z,y)\}\, e\{-B(y-x,z)\}\, e_{(-z,y-x)}\\
&= \sum_{z\in M} e\{B(z,x)\}\, e_{(-z,y-x)} = \sum_{z\in M} e\{B(-z,x)\}\, e_{(z,y-x)}.
\end{aligned}$$

To obtain the last identity we did the substitution $z \mapsto -z$ in the previous sum.

Now we apply the S^*-action in (2.11). The claimed identity holds true for S^* also, since for any $x, y \in M$, similarly we have

$$\kappa S^* e_{x\oplus y} = \frac{\sigma(\underline{M}^{-1} + \underline{M})}{|M|} \sum_{z,y'\in M} e\big\{B(y', z-y)\big\} \sum_{x'\in M} e\big\{B(x', x)\big\}\, e_{(y'-x',z)}$$

$$= \frac{1}{|M|} \sum_{x' \in M} e\left\{B(-x', x)\right\} e_{(x',z)} \sum_{z \in M} \sum_{y' \in M} e\left\{B(y', z - y + x)\right\}$$

$$= \sum_{x' \in M} e\left\{B(-x', x)\right\} e_{(x', y-x)}.$$

To obtain the second identity above we did the substitution $x' \mapsto y' - x'$ and changed the order of summation in the previous sum. Moreover, we also used the fact that $\sigma(\underline{M}^{-1} + \underline{M}) = 1$ which follows from the remark after Definition 1.8 (which says that $\sigma(\underline{M}^{-1} + \underline{M}) = \sigma(\underline{M}^{-1})\sigma(\underline{M}) = \overline{\sigma(\underline{M})}\sigma(\underline{M})$) and Proposition 1.10 (which says that $\sigma(\underline{M})$ has absolute value one). The last identity follows from the fact that the inner sum in the previous identity is 0 unless $z = y - x$, when it equals $|M|$ (see Proposition 1.11). □

Lemma 2.57 *Let $\underline{M}$ be an $\mathcal{O}$-FQM. Then we have*

$$\mathrm{Hom}_{\tilde{\Gamma}}\left(W(\underline{M}), W(\underline{M})\right) \simeq P(\underline{M})^{\tilde{\Gamma}}.$$

Proof This is immediate from Lemmas 2.51 and 2.56. □

Lemma 2.58 *Let $\underline{M} = (M, Q)$ be an $\mathcal{O}$-FQM. For fixed $v \in M \times M$, the map $\chi_v : \mathrm{Stab}(v) \to \mu_\infty$, $A \mapsto f_A(v)$ defines a group homomorphism.*

Proof This is immediate from Lemma 2.53. □

Proof of Theorem 2.7 The number of irreducible $\tilde{\Gamma}$-submodules of the space $W(\underline{M})$ is bounded by the dimension of the space $\mathrm{Hom}_{\tilde{\Gamma}}\left(W(\underline{M}), W(\underline{M})\right)$ (see Proposition 2.22), which equals by Lemma 2.57 the dimension of $P(\underline{M})^{\tilde{\Gamma}}$. But we have $P(\underline{M})^{\tilde{\Gamma}} = P(\underline{M})^{\Gamma_{\mathcal{O}/\mathfrak{l}}}$ (see the first remark after Definition 2.55 and Proposition 2.2). It enough to show that the dimension of the space $P(\underline{M})^{\Gamma_{\mathcal{O}/\mathfrak{l}}}$ equals the upper bound given in the statement of the theorem.

Let v_i $(1 \le i \le m)$ be a set of representatives for $\left(M \times M\right)/\Gamma_{\mathcal{O}/\mathfrak{l}}$. We claim that the space $P(\underline{M})^{\Gamma_{\mathcal{O}/\mathfrak{l}}}$ has as basis the elements

$$L_i := \frac{1}{|\Gamma_{\mathcal{O}/\mathfrak{l}}|} \sum_{A \in \Gamma_{\mathcal{O}/\mathfrak{l}}} \overline{f_{A^{-1}}(v_i)} e_{v_i A^{-1}} \quad (1 \le i \le m)$$

unless they are zero. The space $P(\underline{M})^{\Gamma_{\mathcal{O}/\mathfrak{l}}}$ is spanned by the operators L_i (see Proposition 2.15), and obviously these elements are linearly independent whenever they are nonzero.

First we show that whenever v_i and v_j lie in the same orbit, L_i and L_j differ by a constant. Write $v_j = v_i B^{-1}$ for some $B \in \Gamma_{\mathcal{O}/\mathfrak{l}}$. But the claim holds true, since we have

$$L_j = \frac{1}{|\Gamma_{\mathcal{O}/\mathfrak{l}}|} \sum_{A \in \Gamma_{\mathcal{O}/\mathfrak{l}}} \overline{f_{A^{-1}}(v_i B^{-1})} e_{(v_i B^{-1})A^{-1}}$$

$$= \frac{1}{\overline{f_{B^{-1}}(v_i)}|\Gamma_{\mathcal{O}/\mathfrak{l}}|} \sum_{A \in \Gamma_{\mathcal{O}/\mathfrak{l}}} \overline{f_{(AB)^{-1}}(v_i)} e_{v_i (AB)^{-1}} = \frac{1}{\overline{f_{B^{-1}}(v_i)}} L_i.$$

For the second identity we used Lemma 2.53, and for the last identity we did the substitution $A \mapsto AB^{-1}$ in the previous sum.

Next we determine when the operators L_i are equal to zero. We write

$$L_i = \frac{1}{|\Gamma_{\mathcal{O}/\mathfrak{l}}|} \sum_{A\,\mathrm{Stab}(v_i) \in \Gamma_{\mathcal{O}/\mathfrak{l}}/\,\mathrm{Stab}(v_i)} \sum_{B \in \mathrm{Stab}(v_i)} \overline{f_{(AB)^{-1}}(v_i)} e_{v_i (AB)^{-1}}$$

$$= \frac{1}{|\Gamma_{\mathcal{O}/\mathfrak{l}}|} \sum_{A\,\mathrm{Stab}(v_i) \in \Gamma_{\mathcal{O}/\mathfrak{l}}/\,\mathrm{Stab}(v_i)} e_{v_i A^{-1}} \sum_{B \in \mathrm{Stab}(v_i)} \overline{f_{(AB)^{-1}}(v_i)}.$$

For the second identity we used $v_i B^{-1} = v_i$ which follows from the fact that B is an element of $\mathrm{Stab}(v_i)$. Since the elements $e_{v_i A^{-1}}$ ($A \in \Gamma_{\mathcal{O}/\mathfrak{l}}$) are linearly independent, the operators L_i are equal to zero if for all $A \in \Gamma_{\mathcal{O}/\mathfrak{l}}$, we have

$$\sum_{B \in \mathrm{Stab}(v_i)} \overline{f_{(AB)^{-1}}(v_i)} = 0.$$

However, from Lemma 2.53, and the fact that $v_i B^{-1} = v_i$, we have the identity $f_{(AB)^{-1}}(v_i) = f_{B^{-1}}(v_i) f_{A^{-1}}(v_i)$. Since $f_{A^{-1}}(v_i)$ being a root of unity can never be zero, hence we have

$$\sum_{B \in \mathrm{Stab}(v_i)} \overline{f_{B^{-1}}(v_i)} = 0.$$

But if $L_i = 0$, then χ_{v_i} must be nontrivial (see Lemma 2.58 for χ_{v_i}).

As a consequence, we obtain that the space $P(\underline{M})^{\Gamma_{\mathcal{O}/\mathfrak{l}}}$ has basis the operators L_i for which the characters χ_{v_i} are trivial. Therefore, the dimension of the space $P(\underline{M})^{\Gamma_{\mathcal{O}/\mathfrak{l}}}$ equals the number of elements of $(M \times M)'/\Gamma_{\mathcal{O}/\mathfrak{l}}$. □

For the proof of Corollary 2.47 we need a lemma. Recall that the annihilator and the modified level of an $\mathcal{O}$-CM $\underline{M}$ of level $\mathfrak{l}$ equals $\mathfrak{l}(2,\mathfrak{l})^{-1}$ and $\mathfrak{l}(2,\mathfrak{l})^{-2}$, respectively.

Lemma 2.59 *Let $\mathfrak{a}$ be a nonzero integral $\mathcal{O}$-ideal, let $R := \mathcal{O}/\mathfrak{a}$ and let $\mathfrak{I}$ stand for the set of integral ideals of R. We define*

$$I : (R \times R)/\Gamma_R \to \mathfrak{I}, \quad [\alpha, \beta] \mapsto \alpha R + \beta R.$$

Here Γ_R acts on $R \times R$ via formal multiplication of row vectors in $R \times R$ with matrices in Γ_R. Moreover, $[\alpha, \beta]$ stands for the orbit of (α, β) of under this action. Then the map I defines a bijection.

Proof First we show that I is well-defined. Let $v = (\alpha, \beta)$ and $w = (\alpha', \beta')$ in $R \times R$. We need to show that if v, w lie in the same orbit, then $I([v]) = I([w])$. Suppose v and w lie in the same orbit i.e. $w = vA$ for some $A = \begin{pmatrix} \eta & \xi \\ \gamma & \delta \end{pmatrix} \in \Gamma_R$. Then we have

$$I([w]) = I([vA]) = (\alpha\eta + \beta\gamma)R + (\alpha\xi + \beta\delta)R = (\eta + \xi)\alpha R + (\gamma + \delta)\beta R.$$

Hence, $I([w]) \subseteq I([v])$. On the other hand, we have

$$I([v]) = I([wA^{-1}]) = (\alpha'\delta - \beta'\gamma)R + (-\alpha'\xi + \beta'\eta)R = (\delta - \xi)\alpha' R + (\eta - \gamma)\beta' R.$$

Similarly, we have $I([v]) \subseteq I([w])$ which proves the well-definedness.

The surjectivity of I follows from Lemma 1.25. Next we prove the injectivity. Suppose $I\big([v]\big) = I\big([w]\big)$. From Lemma 1.30 we have that every orbit contains an element whose first entry equals zero. Suppose $(0, \gamma_1)$ is contained in $[v]$, and $(0, \gamma_2)$ is contained in $[w]$. Hence, we have $\gamma_1 R = \gamma_2 R$ (by the assumption). By applying Lemma 1.26, we obtain $\gamma_1 = \varepsilon\gamma_2$ for some $\varepsilon \in R^*$. Hence, $(0, \gamma_1) = (0, \gamma_2)\begin{pmatrix} \varepsilon^{-1} & 0 \\ 0 & \varepsilon \end{pmatrix}$, i.e $[v] = [w]$, which proves that I is an injection. □

Proof of Corollary 2.47 Let $\mathfrak{l}$ and $\mathfrak{a}$ be the level and annihilator of $\underline{M}$, respectively. We set $R := \mathcal{O}/\mathfrak{a}$. From Theorem 2.7 we know that the number of irreducible $\tilde{\Gamma}$-submodules of $W(\underline{M})$ is less than or equal to the number of elements of $\big(R \times R\big)'/\Gamma_{\mathcal{O}/\mathfrak{l}}$. But we have $\big(R \times R\big)'/\Gamma_{\mathcal{O}/\mathfrak{l}} = \big(R \times R\big)'/\Gamma_R =: U$ (see Proposition 2.2 and the first remark after Definition 2.55). It is enough to prove the identity $|U| = \sigma_0(\mathfrak{m})$. Let I be the bijection in Lemma 2.59, and let $\mathfrak{I}$ be the set of integral $\mathcal{O}$-ideals of R. If can show that

$$I(U) = \{(x + \mathfrak{a})R \subseteq \mathfrak{I} : (2, \mathfrak{l})|x\}, \tag{2.29}$$

then the claimed identity holds true. Indeed, since $(x + \mathfrak{a})R$ being an ideal of R must contain $\mathfrak{a}$, i.e. we have $x|\mathfrak{a}$, and since $\mathfrak{a}(2, \mathfrak{l})^{-1} = \mathfrak{m}$. By Theorem 2.7 we have

$$U = \{[v] \in \big(R \times R\big)/\Gamma_R : \chi_v = 1\},$$

where χ_v is as in Lemma 2.58. Let $[v] \in U$. From Lemma 1.30 we know that $[v]$ contains an element of the form $(0, x + \mathfrak{a})$ for some $x \in \mathcal{O}$. Then, since $\chi_v = 1$, we have $f_A(0, x + \mathfrak{a}) = 1$, for all $A \in \mathrm{Stab}(0, x + \mathfrak{a})$. By a direct computation, we obtain

$$\mathrm{Stab}(0, x + \mathfrak{a}) = \left\{ \begin{pmatrix} a+\mathfrak{a} & b+\mathfrak{a} \\ c+\mathfrak{a} & d+\mathfrak{a} \end{pmatrix} \in \Gamma_R : \mathfrak{a}|cx, \mathfrak{a}|dx - x \right\}$$

and $f_A(0, x+\mathfrak{a}) = e\{-cd\omega x^2\}$. To prove (2.29), we need to show that the following holds true:

$$\forall A = \begin{pmatrix} a+\mathfrak{a} & b+\mathfrak{a} \\ c+\mathfrak{a} & d+\mathfrak{a} \end{pmatrix} \in \mathrm{Stab}(0, x+\mathfrak{a}), \quad \mathfrak{l}|cdx^2 \quad \text{if and only if } (2, \mathfrak{l})|x. \tag{2.30}$$

Suppose first of all that $(2, \mathfrak{l})|x$ and $A = \begin{pmatrix} a+\mathfrak{a} & b+\mathfrak{a} \\ c+\mathfrak{a} & d+\mathfrak{a} \end{pmatrix}$ in $\mathrm{Stab}(0, x+\mathfrak{a})$. Hence, we have $\mathfrak{a}|cdx$. Therefore, $\mathfrak{l} = \mathfrak{a}(2, \mathfrak{l})|cdx^2$. (Recall here that $\mathfrak{a} = \mathfrak{l}(2, \mathfrak{l})^{-1}$.)

Suppose now that the left hand side of (2.30) holds true. Let $\mathfrak{c}$ be an integral $\mathcal{O}$-ideal which lies in the inverse ideal class of $\mathfrak{a}$ which is relatively prime to $\mathfrak{l}$. Then $\eta\mathcal{O} = \mathfrak{c}\mathfrak{a}$ for some $\eta \in K$. We take $c = x^{-1}\eta$, $d = 1 + x^{-1}\eta$. Then we have $cdx^2 = x^{-1}\eta(1 + x^{-1}\eta)x^2 = \eta^2 + \eta x$. By the assumption $\mathfrak{l}|cdx^2$, we have that $\mathfrak{l}$ divides $\mathfrak{c}^2\mathfrak{a}^2 + \mathfrak{c}\mathfrak{a}x$, i.e. $\mathfrak{l}\mathfrak{a}^{-1}$ divides $\mathfrak{c}^2\mathfrak{a} + \mathfrak{c}x$. But $\mathfrak{c}$ is chosen so that it is relatively prime to $\mathfrak{l}$, hence $\mathfrak{l}\mathfrak{a}^{-1}$ divides $\mathfrak{c}\mathfrak{a} + x$. But this implies that $\mathfrak{l}\mathfrak{a}^{-1}$, which equals $(2, \mathfrak{l})$, divides x, since $(2, \mathfrak{l})$ also divides $\mathfrak{c}\mathfrak{a}$. Therefore the identity (2.30) holds true, which proves finally the corollary. □

2.6.2 *The Second Approach*

Let $\underline{M}$ be an $\mathcal{O}$-FQM with level $\mathfrak{l}$. In this subsection we use some tools which we already introduced in the previous subsection. Namely, we use the action of $\Gamma_{\mathcal{O}/\mathfrak{l}}$ on $M \times M$ (as given in Lemma 2.52) and the remark afterwards, and we also use the function $f_A(v)$ ($v \in M \times M$) attached to an element A of $\Gamma_{\mathcal{O}/\mathfrak{l}}$ (as given in Lemma 2.53).

In this subsection we give another proof of Theorem 2.7, and we shall prove Theorem 2.8. For the proofs of these theorems we need again several lemmas.

Lemma 2.60 *Let $\underline{M} = (M, Q)$ be an $\mathcal{O}$-FQM with bilinear form B. The space $\mathbb{C}[M]$ is a projective $M \times M$-module via*

$$((x, y), e_z) \mapsto (x, y)e_z := e\{-B(z, y)\}\, e_{x+z}.$$

More precisely, one has $v(we_z) = \lambda(v, w)(v + w)e_z$, where

$$\lambda(v, w) = e\{-B(x', y)\} \qquad (v = (x, y),\ w = (x', y')). \tag{2.31}$$

Proof Let $v = (x, y)$ and $w = (x', y')$ be in $M \times M$ and $z \in M$. Then we have

$$v(we_z) = e\{-B(z, y')\}\, e\{-B(x' + z, y)\}\, e_{x+x'+z} = \lambda(v, w)(v + w)e_z,$$

where $\lambda(v, w) = e\{-B(x', y)\}$. □

Definition 2.61 Let $\underline{M} = (M, Q)$ be an $\mathcal{O}$-FQM with associated bilinear form B. Let l be the level of the finite quadratic $\mathbb{Z}$-module $\mathrm{Tr}(\underline{M})$ (see Proposition 1.3). We define

$$H(\underline{M}) := \{(v, \xi) : v \in M \times M, \xi \in \mu_l\}$$

with the operation

$$(v, \xi) \cdot (w, \xi') = \big(v + w, \xi\xi'\lambda(v, w)\big),$$

where $\lambda(v, w)$ denotes the cocycle (2.31). This group is called the *Heisenberg group associated to* $\underline{M}$. In the sequel, we write (x, y, ξ) instead of $((x, y), \xi)$ for the elements of $H(\underline{M})$.

Remark From Lemma 2.60 and Proposition 2.25 we see that $H(\underline{M})$ is indeed a group, more precisely, a central extension of $M \times M$ by μ_l.

Lemma 2.62 *Let* $\underline{M} = (M, Q)$ *be an* $\mathcal{O}$-*FQM. The space* $\mathbb{C}[M]$ *is an* $H(\underline{M})$-*module via*

$$\big((v, \xi), e_z\big) \mapsto (v, \xi)e_z := \xi \cdot ve_z.$$

For the action of $M \times M$ *on* $\mathbb{C}[M]$, *we refer the reader to Lemma 2.60.*

Proof By Proposition 2.26 and Lemma 2.60, it follows that $\mathbb{C}[M]$ is an $H(\underline{M})$-module. □

Lemma 2.63 *Let* $\underline{M} = (M, Q)$ *be an* $\mathcal{O}$-*FQM. The character* $\chi_{\mathbb{C}[M]}$ *of the* $H(\underline{M})$-*module* $\mathbb{C}[M]$ *satisfies*

$$\chi_{\mathbb{C}[M]}(v, \xi) = \begin{cases} 0 & \text{if } v \neq 0 \\ \xi|M| & \text{otherwise.} \end{cases}$$

Proof From Lemma 2.62, $\mathbb{C}[M]$ is an $H(\underline{M})$-module. Let B be the bilinear form of $\underline{M}$ and $\big(v = (x, y), \xi\big) \in H(\underline{M})$. The following identity proves the claimed identity

$$\mathrm{tr}\,\big((v, \xi), \mathbb{C}[M]\big) = \begin{cases} 0 & \text{if } x \neq 0 \\ \xi \sum_{z \in M} e\,\{-B(z, y)\} & \text{otherwise,} \end{cases}$$

since the sum above is zero unless $y = 0$, when it equals $|M|$ (see Proposition 1.11). □

Lemma 2.64 *Let* $\underline{M} = (M, Q)$ *be an* $\mathcal{O}$-*FQM. The space* $\mathbb{C}[M]$ *is an irreducible* $H(\underline{M})$-*module.*

Proof Using Lemma 2.63, we have

$$\frac{1}{H(\underline{M})}\sum_{(v,\xi)\in H(\underline{M})}|\chi_{\mathbb{C}[M]}(v,\xi)|^2 = \frac{1}{l|M|^2}\sum_{\xi\in\mu_l}|M|^2|\xi|^2 = \frac{1}{l|M|^2}|M|^2 l = 1$$

which proves the lemma (using [FH91, Cor. 2.15]). □

Lemma 2.65 *Let* $\underline{M} = (M, Q)$ *be an* $\mathcal{O}$*-FQM with level* $\mathfrak{l}$*. The group* $\Gamma_{\mathcal{O}/\mathfrak{l}}$ *acts from the right on* $H(\underline{M})$ *via*

$$\big((v,\xi), A\big) \mapsto (v,\xi)^A := \big(vA, \xi f_A(v)\big),$$

where $f_A(v)$ *is as in Lemma 2.53.*

Remark Note that the above map commutes with the embedding ι and the canonical projection π given in the following exact sequence

$$1 \to \mu_l \xrightarrow{\iota} H(\underline{M}) \xrightarrow{\pi} M \times M \to 1.$$

Proof of Lemma 2.65 It is enough to show that for fixed A in $\Gamma_{\mathcal{O}/\mathfrak{l}}$, the map $h \mapsto h^A$ defines a group homomorphism of $H(\underline{M})$, since Lemma 2.53 and the remark after Lemma 2.52 ensure the fact that the map in the statement of the lemma satisfies the axioms of an action.

Let $A = \left(\begin{smallmatrix} a+\mathfrak{l} & b+\mathfrak{l} \\ c+\mathfrak{l} & d+\mathfrak{l}\end{smallmatrix}\right)$ be in $\Gamma_{\mathcal{O}/\mathfrak{l}}$ and let $h = (v, \xi)$ and $h' = (w, \xi')$ be in $H(\underline{M})$. We have

$$h^A \cdot h'^A = \big(vA + wA, \xi\xi' f_A(v) f_A(w)\lambda(vA, wA)\big).$$

On the other hand, we have

$$(h \cdot h')^A = \big(vA + wA, \xi\xi'\lambda(v, w) f_A(v + w)\big).$$

Hence, it remains to show that the following identity holds true

$$\lambda(v, w) f_A(v + w) = \lambda(vA, wA) f_A(v) f_A(w).$$

Calculating both sides separately and inserting the following values

$$\lambda(vA, wA) = e\left\{-B(ax' + cy', bx + dy)\right\}, \quad \lambda(v, w) = e\left\{-B(x', y)\right\},$$

we see that the following identity proves the assertion:

$$\begin{aligned}
&\lambda(vA, wA) f_A(v) f_A(w) \\
&= e\left\{-B(ax' + cy', bx + dy)\right\} e\left\{-\big(abQ(x) + bcB(x, y) + cdQ(y)\big)\right\} \times \\
&\quad \times e\left\{-\big(abQ(x') + bcB(x', y') + cdQ(y')\big)\right\}
\end{aligned}$$

$$
\begin{aligned}
&= e\left\{-\big(abQ(x+x') + cdQ(y+y')\big)\right\} \times \\
&\quad \times e\left\{-\big(bcB(x, y+y') + bcB(x', y+y') + B(x', y)\big)\right\} \\
&= e\left\{-\big(abQ(x+x') + bcB(x+x', y+y') + cdQ(y+y')\big)\right\} \times \\
&\quad \times e\left\{-B(x', y)\right\} = \lambda(v, w) f_A(v+w).
\end{aligned}
$$

For the third identity we used $ad - bc \equiv 1 \bmod \mathfrak{l}$, and the fact that $lB = 0$ for any $l \in \mathfrak{l}$. □

Definition 2.66 Let $\underline{M}$ be an $\mathcal{O}$-FQM with level $\mathfrak{l}$. Using the action of $\Gamma_{\mathcal{O}/\mathfrak{l}}$ on $H(\underline{M})$ from Lemma 2.65, we define

$$J(\underline{M}) := \Gamma_{\mathcal{O}/\mathfrak{l}} \ltimes H(\underline{M}).$$

We call $J(\underline{M})$ as the *Jacobi group associated to* $\underline{M}$.

Remark The group operation in $J(\underline{M})$ is given by

$$(A, h) \cdot (B, h') := (AB, h^B \cdot h').$$

The fact that $J(\underline{M})$ becomes a group with this operation is a well-known fact in basic algebra.

More explicitly, for $h = (v, \xi)$ and $h' = (w, \xi')$, the above operation is given by

$$\big(A, (v, \xi)\big) \cdot \big(B, (w, \xi')\big) = \big(AB, ((vB) + w, \xi\xi' f_B(v)\lambda((vB), w))\big).$$

Remark Henceforth, for the elements of $J(\underline{M})$, we use (A, v, ξ) instead of $(A, (v, \xi))$. We view $\Gamma_{\mathcal{O}/\mathfrak{l}}$ (where $\mathfrak{l}$ is the level of $\underline{M}$) and $H(\underline{M})$ as subgroups of $J(\underline{M})$ via the maps $A \mapsto (A, 1)$ and $h \mapsto (1, h)$, respectively. Moreover, via the map $\alpha \mapsto (\pi(\alpha), 1)$, the group $\tilde{\Gamma}$ can be viewed as a subgroup of $J(\underline{M})$, where π is the epimorphism from $\tilde{\Gamma}$ onto $\Gamma_{\mathcal{O}/\mathfrak{l}}$ explained in the first remark after Definition 2.55.

Lemma 2.67 *Let $\underline{M} = (M, Q)$ be an $\mathcal{O}$-FQM with level $\mathfrak{l}$. For $h = (v, \xi) \in H(\underline{M})$ and $A \in \Gamma_{\mathcal{O}/\mathfrak{l}}$, we have*

$$AhA^{-1} = \big(1, vA^{-1}, \xi f_{A^{-1}}(v)\big).$$

Proof The following identity proves the claimed identity:

$$
\begin{aligned}
AhA^{-1} &= (A, 0, 1)(1, v, \xi)(A^{-1}, 0, 1) = (A, 0, 1)\big(A^{-1}, vA^{-1}, \xi f_{A^{-1}}(v)\big) \\
&= \big(1, vA^{-1}, \xi f_{A^{-1}}(v)\big).
\end{aligned}
$$

□

Lemma 2.68 *Let $\underline{M} = (M, Q)$ be an $\mathcal{O}$-FQM with level $\mathfrak{l}$. For fixed $A \in \Gamma_{\mathcal{O}/\mathfrak{l}}$, we define*

$$\sigma_A : H(\underline{M}) \to H(\underline{M}), \quad \sigma_A(h) = AhA^{-1}.$$

If σ is the representation afforded by the $H(\underline{M})$-module $\mathbb{C}[M]$ (see Lemma 2.62), then the representations σ and $\sigma \circ \sigma_A$ of $H(\underline{M})$ are equivalent.

Proof First note from Lemma 2.67 that AhA^{-1} lies in $H(\underline{M})$ for any h in $H(\underline{M})$. It is easy to see that $\sigma \circ \sigma_A$ defines a representation of $H(\underline{M})$. Using Proposition 2.13, it suffices to show that the traces of σ and $\sigma \circ \sigma_A$ are equal. Let B be the bilinear form of $\underline{M}$. Write $A = \left(\begin{smallmatrix} a+\mathfrak{l} & b+\mathfrak{l} \\ c+\mathfrak{l} & d+\mathfrak{l} \end{smallmatrix}\right)$ and let $h = \left(v = (x, y), \xi\right)$ be in $H(\underline{M})$. The trace of $\sigma \circ \sigma_A$ becomes

$$\operatorname{tr}\left(AhA^{-1}, \mathbb{C}[M]\right) = \begin{cases} 0 & \text{if } dx - cy \neq 0 \\ \xi f_{A^{-1}}(v) \sum_{z \in M} e\left\{-B(z, -bx + ay)\right\} & \text{otherwise.} \end{cases}$$

Here we used Lemma 2.67 and the action in Lemma 2.62, also the identity $vA^{-1} = (dx - cy, -bx + ay)$ (see Lemma 2.52 and the remark afterwards).

We use Proposition 1.11 to evaluate the above sum. We obtain that it is zero unless $ay = bx$, when it equals $|M|$. Therefore, we recognize that this value coincides with the trace of σ in Lemma 2.63. □

Lemma 2.69 *Let $\underline{M} = (M, Q)$ be an $\mathcal{O}$-FQM with level $\mathfrak{l}$ and σ be the representation afforded by the $H(\underline{M})$-module $\mathbb{C}[M]$ (see Lemma 2.62). For each $A \in \Gamma_{\mathcal{O}/\mathfrak{l}}$, there exists an (up to multiplication by a constant) unique $\delta(A) \in \mathrm{GL}(\mathbb{C}[M])$ such that the following holds true:*

$$\delta(A)\sigma(h)\delta(A)^{-1} = \sigma(AhA^{-1}) \quad (h \in H(\underline{M})). \tag{2.32}$$

Proof Let $A \in \Gamma_{\mathcal{O}/\mathfrak{l}}$. By Lemma 2.68 we know that $\sigma \circ \sigma_A$ and σ are equivalent to each other. Hence, there exists an element $\delta(A)$ of $\mathrm{GL}(\mathbb{C}[M])$ such that (2.32) holds true. It remains to show that $\delta(A)$ is unique up to multiplication by a constant. Assume there exists $\gamma(A) \in \mathrm{GL}(\mathbb{C}[M])$ which satisfies also (2.32). Let $h \in H(\underline{M})$. We then have

$$\gamma^{-1}\delta(A)\sigma(h)(\gamma^{-1}\delta)^{-1}(A) = \sigma(h).$$

We denote $\gamma^{-1}(A)\delta(A)$ by $\varphi(A)$. Using the above identity we obviously have $\varphi(A)(hv) = h\varphi(A)(v)$ for any $v \in \mathbb{C}[M]$. But this implies that $\varphi(A)$ defines an $H(\underline{M})$-linear map on $\mathbb{C}[M]$. Since from Lemma 2.64 we know that $\mathbb{C}[M]$ is an irreducible $H(\underline{M})$-module, the result follows from Schur's Lemma (see e.g [FH91, Lem. 1.7]). □

Lemma 2.70 *Let $\underline{M} = (M, Q)$ be an $\mathcal{O}$-FQM with level $\mathfrak{l}$. For A an element of* $\mathrm{SL}(2, \mathcal{O})$*, let $\delta(A)$ be an element of* $\mathrm{GL}(\mathbb{C}[M])$ *satisfying* (2.32). *The map $A \mapsto \delta(A)$ defines a projective representation of $\Gamma_{\mathcal{O}/\mathfrak{l}}$.*

Proof Let A and B be in $\Gamma_{\mathcal{O}/\mathfrak{l}}$. By assumption, $\delta(AB)$ satisfies (2.32). It is enough to show that $\delta(A)\delta(B)$ also satisfies the same identity, since then by proceeding as in the proof of Lemma 2.69, the statement of the lemma holds true. But we have

$$\begin{aligned}\delta(A)\delta(B)\sigma(h)(\delta(A)\delta(B))^{-1} &= \delta(A)\sigma\big(BhB^{-1}\big)\delta(A)^{-1} = \sigma\big(ABhB^{-1}A^{-1}\big)\\ &= \sigma(ABh(AB)^{-1}).\end{aligned}$$

Here we used the assumption that $\delta(A)$ and $\delta(B)$ satisfy (2.32). □

Proof of Theorem 2.8 For $z \in M$ and $b \in \mathcal{O}$, we define

$$L(T_b)e_z := e\{bQ(z)\}\, e_z.$$

Let π be the epimorphism in Lemma 1.29 from Γ onto $\Gamma_{\mathcal{O}/\mathfrak{l}}$. If we can show that the operators $L(T_b)$ and $\delta(\pi(T_b))$ differ by a constant, then multiplying $\delta(\pi(T_b))$ with a suitable constant so that it satisfies (i) and using Lemma 2.70, we see that part (i) of theorem holds true. To show that these operators differ by a constant it is enough to show that $L(T_b)$ also satisfies (2.32) (see the proof of Lemma 2.69). But for any $h = (x, y, \xi) \in H(\underline{M})$, we have

$$\begin{aligned}L(T_b)\sigma(h)L(T_b)^{-1}e_z &= \xi e\{bQ(x) + B(z, bx - y)\}\, e_{z+x}\\ &= \xi f_{T_b^{-1}}(v) e\{-B(-bx + y, z)\}\, e_{z+x}\\ &= \sigma\big(\pi(T_b)h\pi(T_b)^{-1}\big)e_z.\end{aligned}$$

To obtain the first identity we used $L(T_b)^{-1}e_z = e\{-bQ(z)\}\, e_z$ and the action in Lemma 2.62. For the second identity we used Lemmas 2.67 and 2.62.

For $z \in M$, we set

$$L(S)e_z := \sigma(\underline{M})\frac{1}{\sqrt{|M|}}\sum_{z' \in M} e\{-B(z', z)\}\, e_{z'}.$$

We show that the operators $L(S)$ and $\delta(\pi(S))$ differ by a constant. Then, proceeding as in the previous case, we obtain that part (ii) of theorem also holds true. But for any $h = (x, y, \xi) \in H(\underline{M})$, we have

$$\begin{aligned}L(S)\sigma(h)L(S)^{-1}e_z = \xi\frac{1}{|M|}\sum_{z'' \in M} e\{-B(x, z'')\}\, e_{z''} \times\\ \times \sum_{z' \in M} e\{-B(z', y - z + z'')\}\end{aligned}$$

$$= \xi e\{B(x, y-z)\} e_{z-y} = \xi f_{S^{-1}}(v) e\{-B(z, x)\} e_{z-y}$$
$$= \sigma(\pi(S) h \pi(S)^{-1}) e_z.$$

For the first identity we used $L(S)^{-1}e_z = \overline{\sigma(\underline{M})}\frac{1}{\sqrt{|M|}}\sum_{z' \in M} e\{B(z', z)\} e_{z'}$ and the action in Lemma 2.62. Moreover, we also used the fact that $\sigma(\underline{M})$ has absolute value one (see Proposition 1.10). The second identity follows from the fact that the sum in the previous identity is zero unless $z'' = y - z$, when it equals $|M|$ (see Proposition 1.11). For the third identity we used Lemmas 2.67 and 2.62. This proves the theorem. □

Remark Let $\underline{M} = (M, Q)$ be an $\mathcal{O}$-FQM with level $\mathfrak{l}$. By (2.3) and also Lemma 2.70 we have that $\Gamma_{\mathcal{O}/\mathfrak{l}}$ acts on $\mathrm{Hom}(\mathbb{C}[M], \mathbb{C}[M])$ via

$$(A, \lambda) \to {}^A\lambda, \quad {}^A\lambda(v) = \delta(A)\lambda\big(\delta(A)^{-1}(v)\big).$$

Here $\delta(A)$ is any element of $\mathrm{GL}(\mathbb{C}[M])$ which satisfies (2.32). Using the first remark after Definition 2.55 and Proposition 2.2 we have that $\tilde{\Gamma}$ also acts on the space $\mathrm{Hom}(\mathbb{C}[M], \mathbb{C}[M])$.

On the other hand, since $W(\underline{M})$ is a $\tilde{\Gamma}$-module (see Sect. 2.2), the group $\tilde{\Gamma}$ acts on $\mathrm{Hom}(W(\underline{M}), W(\underline{M}))$ via (see (2.3))

$$(\alpha, \lambda) \to {}^\alpha\lambda, \quad {}^\alpha\lambda(v) = \rho(\alpha)\lambda\big(\rho(\alpha)^{-1}(v)\big),$$

where ρ is the representation afforded by $\tilde{\Gamma}$-module $W(\underline{M})$. But from Theorem 2.8 we know that ρ and δ differ only by a constant. Therefore, the action of $\tilde{\Gamma}$ on the spaces $\mathrm{Hom}(\mathbb{C}[M], \mathbb{C}[M])$ and $\mathrm{Hom}(W(\underline{M}), W(\underline{M}))$ coincide.

We can now give the second proof of Theorem 2.7.

Proof of Theorem 2.7 The dimension of $\mathrm{Hom}_{\tilde{\Gamma}}\big(W(\underline{M}), W(\underline{M})\big)$ provides a bound for the number of irreducible $\tilde{\Gamma}$-submodules of $W(\underline{M})$ (see Proposition 2.22). By Proposition 2.21 $\mathrm{Hom}_{\tilde{\Gamma}}\big(W(\underline{M}), W(\underline{M})\big)$ in fact equals $\mathrm{Hom}\big(W(\underline{M}), W(\underline{M})\big)^{\tilde{\Gamma}}$. From the previous remark the latter space equals $\mathrm{Hom}\big(\mathbb{C}[M], \mathbb{C}[M]\big)^{\Gamma_{\mathcal{O}/\mathfrak{l}}}$. Therefore, it is enough to show that the dimension of this latter space equals the upper bound given in the statement of the theorem.

Since $\mathbb{C}[M]$ is an irreducible $H(\underline{M})$-module (see Lemma 2.64), the elements $\sigma(v, 1)$ $(v \in M \times M)$ form a basis for the space $\mathrm{Hom}\big(\mathbb{C}[M], \mathbb{C}[M]\big)$ (see [Ser77, Prop. 10]). Here σ is the representation afforded by the $H(\underline{M})$-module $\mathbb{C}[M]$. By Proposition 2.15 and the action given in the previous remark we have that $\mathrm{Hom}\big(\mathbb{C}[M], \mathbb{C}[M]\big)^{\Gamma_{\mathcal{O}/\mathfrak{l}}}$ is spanned by the operators

$$L(v) := \frac{1}{|\Gamma_{\mathcal{O}/\mathfrak{l}}|} \sum_{A \in \Gamma_{\mathcal{O}/\mathfrak{l}}} \delta(A)\sigma(v, 1)\delta(A)^{-1} \quad (v \in M \times M).$$

We claim that the nonzero $L(v)$, where the v are representatives for $(M \times M)/\Gamma_{\mathcal{O}/\mathfrak{l}}$, form a basis for the space $\mathrm{Hom}\left(\mathbb{C}[M], \mathbb{C}[M]\right)^{\Gamma_{\mathcal{O}/\mathfrak{l}}}$. For that first we need to show that if v and w lie in the same orbit (see Lemma 2.52 and the remark afterwards), then $L(v)$ and $L(w)$ differ by a constant. We write $v = wB^{-1}$ for some $B \in \Gamma_{\mathcal{O}/\mathfrak{l}}$. Then we have

$$L(v) = \frac{1}{|\Gamma_{\mathcal{O}/\mathfrak{l}}|} \sum_{A \in \Gamma_{\mathcal{O}/\mathfrak{l}}} \delta(A)\sigma(wB^{-1}, 1)\delta(A)^{-1}.$$

Since $\delta(A)$ satisfies (2.32), we have

$$\delta(A)\sigma(wB^{-1}, 1)\delta(A)^{-1} = \sigma\big(A(wB^{-1}, 1)A^{-1}\big).$$

We know $A(wB^{-1}, 1)A^{-1} = \big(1, wB^{-1}A^{-1}, f_{A^{-1}}(wB^{-1})\big)$ from Lemma 2.67 and also $f_{A^{-1}}(wB^{-1}) = f_{(AB)^{-1}}(w)/f_{B^{-1}}(w)$ from Lemma 2.53. Now inserting these values to the above identity gives

$$\delta(A)\sigma(wB^{-1}, 1)\delta(A)^{-1} = \frac{1}{f_{B^{-1}}(w)}\delta(AB)\sigma(w, 1)\delta(AB)^{-1}.$$

Inserting this quantity to the last sum and doing the substitution $A \mapsto AB^{-1}$ proves the assertion. Next we determine when $L(v) = 0$. We write

$$L(v) = \frac{1}{|\Gamma_{\mathcal{O}/\mathfrak{l}}|} \sum_{B\,\mathrm{Stab}(v) \in \Gamma_{\mathcal{O}/\mathfrak{l}}/\,\mathrm{Stab}(v)} \; \sum_{A \in \mathrm{Stab}(v)} \delta(AB)\sigma(v, 1)\delta(AB)^{-1}.$$

We have the following identity

$$\begin{aligned}\delta(AB)\sigma(v, 1)\delta(AB)^{-1} &= \sigma\big(AB(v, 1)(AB)^{-1}\big) = \sigma\big(1, (vB), f_A(v) f_B(v)\big) \\ &= \sigma((vB)) f_B(v) f_A(v).\end{aligned}$$

The first identity follows since $\delta(AB)$ satisfies (2.32). The second identity follows from Lemmas 2.67, 2.53 and the fact that $A \in \mathrm{Stab}(v)$. The last identity is implied by the action in Lemma 2.62. Therefore, we obtain

$$L(v) = \frac{1}{|\Gamma_{\mathcal{O}/\mathfrak{l}}|} \sum_{B\,\mathrm{Stab}(v) \in \Gamma_{\mathcal{O}/\mathfrak{l}}/\,\mathrm{Stab}(v)} \sigma((vB)) f_B(v) \sum_{A \in \mathrm{Stab}(v)} f_A(v).$$

Clearly, $L(v) = 0$ if and only if the inner sum equals zero. But the inner sum equals zero if and only if χ_v is nontrivial (see Lemma 2.58 for χ_v).

Consequently, the space Hom $\left(\mathbb{C}[M], \mathbb{C}[M]\right)^{\Gamma_{\mathcal{O}/\mathfrak{l}}}$ has as basis the operators $L(v)$ for which the characters χ_v are trivial. Therefore the dimension of the space Hom $\left(\mathbb{C}[M], \mathbb{C}[M]\right)^{\Gamma_{\mathcal{O}/\mathfrak{l}}}$ equals the number of elements in $(M \times M)'/\Gamma_{\mathcal{O}/\mathfrak{l}}$. □

Chapter 3
Jacobi Forms over Totally Real Number Fields

From this chapter on, the number field K is assumed to be totally real. This restriction is necessary for guaranteeing the holomorphicity of *Jacobi forms*. As before, we shall simply write $\mathscr{O}$, $\mathfrak{d}$ for the ring of integers and different of K, respectively. Furthermore, we shall use $\Gamma = \mathrm{SL}(2, \mathscr{O})$, and we shall write $\tilde{\Gamma}$ for the *metaplectic cover* of $\mathrm{SL}(2, \mathscr{O})$ which will be defined in Sect. 3.3. In addition, for a subring R of K, we shall denote by Γ_R the group $\mathrm{SL}(2, R)$ and by $\tilde{\Gamma}_R$ the metaplectic cover of Γ_R.

In the present chapter we shall develop a theory for *Jacobi forms over number fields*. In particular, we shall see that there is a one-to-one correspondence between the spaces of Jacobi forms and certain spaces of vector-valued Hilbert modular forms (see Theorem 3.5). As an immediate corollary, we shall deduce that the spaces of Jacobi forms are finite dimensional (see Corollary 3.53). Certain spaces of functions, the *Jacobi theta functions* which can be viewed as modules over $\tilde{\Gamma}$ (see Theorem 3.1), will play an important role in this context. In the next chapter, we shall define a $\tilde{\Gamma}$-module isomorphism between the spaces of Weil representations associated to certain discriminant modules and the spaces of these theta functions. This will be a key step for the explicit description of the singular Jacobi forms whose index is a rank one $\mathscr{O}$-lattice. We shall also calculate the matrix coefficients of the action of $\tilde{\Gamma}$ on the spaces of Jacobi theta functions (see Theorem 3.1).

In Sect. 3.1, we shall recall or develop those basic facts about integral lattices over number fields (the *$\mathscr{O}$-lattices*) which are crucial for the definition of Jacobi forms. In Sect. 3.2, we shall introduce some basic notations which will help to avoid clumsy notations when dealing with Jacobi forms and Hilbert modular forms. In Sect. 3.3, we shall define the metaplectic cover of $\mathrm{SL}(2, \mathscr{O})$, which will be necessary to include Jacobi forms of half integral weight. In Sect. 3.4, we shall introduce the notions of *Heisenberg groups* and the *Jacobi groups* associated to $\mathscr{O}$-lattices, and we list several results concerning the actions of these groups on the spaces of holomorphic functions, which will be helpful for defining Jacobi forms. In Sect. 3.5,

H. Boylan, *Jacobi Forms, Finite Quadratic Modules and Weil Representations over Number Fields*, Lecture Notes in Mathematics 2130,
DOI 10.1007/978-3-319-12916-7_3

we shall introduce the spaces of Jacobi theta functions associated to $\mathcal{O}$-lattices. Later in the same section, we shall study the spaces of Jacobi theta functions as $\tilde{\Gamma}$-modules, and moreover we shall determine the matrix coefficients of the $\tilde{\Gamma}$-action. In Sect. 3.6, we shall finally define Jacobi forms, and we study their *Fourier developments* and *theta expansions*. In Sect. 3.7, we shall show that the spaces of Jacobi forms are isomorphic to spaces of vector-valued Hilbert modular forms. In particular, we shall be able to prove that a Köcher principle holds true for Jacobi forms and that the spaces of Jacobi forms are finite dimensional.

3.1 $\mathcal{O}$-Lattices

Definition 3.1 An *integral lattice over* $\mathcal{O}$ is a pair $\underline{L} = (L, \beta)$, where L denotes a finitely generated torsion-free $\mathcal{O}$-module, and where $\beta : L \times L \to \mathfrak{d}^{-1}$ is a map which satisfies the following properties:

(i) The map β is $\mathcal{O}$-bilinear and symmetric.
(ii) The map β is non-degenerate (i.e. $\beta(x, L) = \{0\}$ if and only if $x = 0$).

For simplicity, the integral lattices over $\mathcal{O}$ will be named *$\mathcal{O}$-lattices*. We sometimes write shortly $x \in \underline{L}$ for $x \in L$.

Proposition 3.2 *Let* $\underline{L} = (L, \beta)$ *be an* $\mathcal{O}$*-lattice. Then* $\operatorname{Tr}(\underline{L}) := \big(L, \operatorname{tr}_{K/\mathbb{Q}} \circ \beta\big)$ *defines a* $\mathbb{Z}$*-lattice.*

Proof The bilinear form $\operatorname{tr}_{K/\mathbb{Q}} \circ \beta$ is non-degenerate, since the $\mathbb{Q}$-bilinear map $(a, b) \mapsto \operatorname{tr}_{K/\mathbb{Q}}(a, b)$ is non-degenerate ($a, b \in K$). Clearly, $\operatorname{tr}_{K/\mathbb{Q}} \circ \beta$ is $\mathbb{Z}$-bilinear and symmetric, which proves the proposition. □

Let $\underline{L} = (L, \beta)$ be an $\mathcal{O}$-lattice. The $\mathcal{O}$-lattice $\underline{L}' = (L', \beta')$ is called an *$\mathcal{O}$-sublattice of* $\underline{L}$, if L' is an $\mathcal{O}$-submodule of L, and β' is the restriction of β to $L' \times L'$.

For $x \in L^{\#}$, here and in the following, we set $\beta(x) := \frac{1}{2}\beta(x, x)$. If $\beta(x) \in \mathfrak{d}^{-1}$, then $\underline{L}$ is called an *even* $\mathcal{O}$*-lattice*, otherwise it is called an *odd* $\mathcal{O}$*-lattice*. Every odd $\mathcal{O}$-lattice contains an even $\mathcal{O}$-sublattice. Indeed, the map $x \mapsto \beta(x, x) + 2\mathfrak{d}^{-1}$ defines a group homomorphism from L to $\mathfrak{d}^{-1}/2\mathfrak{d}^{-1}$. If $\underline{L}$ is even, then it is the trivial homomorphism. If $\underline{L}$ is odd, then the kernel of this homomorphism is an even $\mathcal{O}$-sublattice of $\underline{L}$.

For $1 \le j \le n = [K : \mathbb{Q}]$, let σ_j be the embeddings of K into $\mathbb{R}$. If $\sigma_j \circ \beta(x, x) > 0$ for all j and all nonzero $x \in L$, then $\underline{L}$ is called *totally positive definite*. Note that the notion of totally positive definite $\mathcal{O}$-lattices is a generalization to number fields of positive and integral $\mathbb{Z}$-lattices.

We say that *there is a homomorphism from* $\underline{L}$ *to* $\underline{L}'$, if there is an $\mathcal{O}$-module homomorphism $\varphi : L \to L'$ which is isometric, i.e. such that $\beta'(\varphi(x), \varphi(y)) = \beta(x, y)$ ($x, y \in L$). Note that every homomorphism φ between totally positive definite $\mathcal{O}$-lattices is injective (indeed, if $\varphi(x) = 0$, then $0 = \beta'\big(\varphi(x), \varphi(y)\big) =$

$\beta(x, y)$ for all y, which implies $x = 0$ since β is non-degenerate). The $\mathcal{O}$-lattices $\underline{L}$ and $\underline{L}'$ are called *isomorphic*, and we write $\underline{L} \simeq \underline{L}'$, if there is an isomorphism between them.

Recall that every torsion-free finitely generated $\mathcal{O}$-module L is isomorphic as an $\mathcal{O}$-module to an $\mathcal{O}$-module of the form $\mathcal{O}^{r-1} \oplus \mathfrak{a}$ for some positive integer r and a fractional $\mathcal{O}$-ideal $\mathfrak{a}$ [FT93, § II.4, Thm. 13(b)]. Moreover, the integer r and the ideal class of $\mathfrak{a}$ are uniquely determined by L [FT93, § II.4, Thm 13(c)]. The integer r is called the *rank of* $\underline{L}$. The ideal class of $\mathfrak{a}$ is called the *Steinitz-invariant of* $\underline{L}$. Clearly, r equals the dimension of the K-vector space $K \otimes_{\mathcal{O}} L$, i.e. $r = \dim_K K \otimes_{\mathcal{O}} L$.

If A is a ring extension of $\mathcal{O}$, we denote the A-bilinear extension of β, namely the bilinear map $A \otimes_{\mathcal{O}} L \times A \otimes_{\mathcal{O}} L \to A \otimes_{\mathcal{O}} \mathfrak{d}^{-1}$, also by β. We use $\beta(x) = \frac{1}{2}\beta(x, x)$ if 2 is invertible in $A \otimes_{\mathcal{O}} \mathfrak{d}^{-1}$. If $\mathfrak{d}^{-1}$ is contained in A, we identify the A-module $A \otimes_{\mathcal{O}} \mathfrak{d}^{-1}$ with A (via the $\mathcal{O}$-linear map induced from the $\mathcal{O}$-bilinear map $(a, d) \mapsto ad$). The *dual of* $\underline{L}$ is defined as

$$L^\# := \{x \in K \otimes_{\mathcal{O}} L : \beta(x, L) \in \mathfrak{d}^{-1}\}.$$

Note that $L^\#$ is again a finitely generated torsion-free $\mathcal{O}$-module, and that we have $(L^\#)^\# = L$ [O'M00, §82F] (loc. cit. the dual of a lattice is defined slightly differently than ours, but it is easy to modify the arguments given loc. cit. so that they extend also to our situation).

Definition 3.3 Let $\underline{L} = (L, \beta)$ be an even $\mathcal{O}$-lattice. We define the *discriminant module of* $\underline{L}$ as the $\mathcal{O}$-FQM:

$$D_{\underline{L}} := \left(L^\#/L, x + L \mapsto \beta(x) + \mathfrak{d}^{-1}\right).$$

Remark Note that, for the well-definedness of the quadratic form $Q : x + L \mapsto \beta(x) + \mathfrak{d}^{-1}$ of $D_{\underline{L}}$ the evenness of $\underline{L}$ is crucial. The non-degeneracy of Q comes from the fact that $(L^\#)^\# = L$.

From [Ebe02, § 1.1] we know that $L^\#/L$ is finite. By the *level*, *modified level* and the *annihilator of an* $\mathcal{O}$*-lattice* $\underline{L}$ we mean the level, modified level and the annihilator of the $\mathcal{O}$-FQM $D_{\underline{L}}$. The reader is referred to Sect. 1.1 for basic notions about finite quadratic $\mathcal{O}$-modules.

Definition 3.4 Let $\underline{L} = (L, \beta)$ be an even $\mathcal{O}$-lattice. For an isotropic submodule U of $L^\#/L$, we define $\underline{L}/U := \left(\pi^{-1}(U), \beta\right)$, where π is the canonical projection from $L^\#$ onto $L^\#/L$.

Remark Note that $\underline{L}/U$ is again an even $\mathcal{O}$-lattice (the non-degeneracy of β on $\pi^{-1}(U) \times \pi^{-1}(U)$ follows from the easily proven fact that $x \mapsto \beta(x, \cdot)$ defines an isomorphism of the K-vector space $K \otimes_{\mathcal{O}} L$ with its dual and that $\pi^{-1}(U)$ contains a basis of this vector space.)

Proposition 3.5 *Let* $\underline{L} = (L, \beta)$ *be an even* $\mathcal{O}$*-lattice and* π *be the canonical projection from* $L^\#$ *to* $L^\#/L$*. The map* $x + \pi^{-1}(U) \mapsto \pi(x) + U$ *defines an isomorphism from* $D_{\underline{L}/U}$ *to* $D_{\underline{L}}/U$*.*

Proof The statement is an obvious consequence of the very definition of $\underline{L}/U$. □

The remaining statements of this section concern $\mathcal{O}$-lattices of rank one and will not be used before the next chapter.

Definition 3.6 Let $\mathfrak{c}$ be a nonzero fractional $\mathcal{O}$-ideal, and ω be a nonzero element of K^* such that $\omega \gg 0$ and $\omega\mathfrak{c}^2 \subseteq \mathfrak{d}^{-1}$. We set

$$(\mathfrak{c}, \omega) := (\mathfrak{c}, (x, y) \mapsto \omega xy). \tag{3.1}$$

Note that, $(\mathfrak{c}, \omega)$ defines a totally positive definite $\mathcal{O}$-lattice of rank one.

Proposition 3.7 *Let* $(\mathfrak{c}, \omega)$ *be as in the above definition, and assume that* $(\mathfrak{c}, \omega)$ *is even. Then the discriminant module of* $(\mathfrak{c}, \omega)$ *is an* $\mathcal{O}$*-CM. Moreover, the annihilator, level and the modified level of* $(\mathfrak{c}, \omega)$ *equal* $\mathfrak{c}^2\omega\mathfrak{d}$, $2\mathfrak{c}^2\omega\mathfrak{d}$ *and* $\mathfrak{c}^2\omega\mathfrak{d}/2$*, respectively. The annihilator and the level of* $\underline{L}$ *are divisible by* 2 *and* 4*, respectively.*

Proof Let $\underline{L} = (L, \beta) = (\mathfrak{c}, \omega)$. Since $L^\# = \{y \in K : \omega y\mathfrak{c} \subseteq \mathfrak{d}^{-1}\} = (\mathfrak{c}\omega\mathfrak{d})^{-1}$, we have

$$D_{\underline{L}} = \left((\mathfrak{c}\omega\mathfrak{d})^{-1}/\mathfrak{c}, x + \mathfrak{c} \mapsto \omega x^2/2 + \mathfrak{d}^{-1}\right).$$

Here note that $\mathfrak{c} \subseteq (\mathfrak{c}\omega\mathfrak{d})^{-1}$, since the $\mathcal{O}$-lattice $(\mathfrak{c}, \omega)$ is integral. Lemma 1.17 implies then that $D_{\underline{L}}$ is an $\mathcal{O}$-CM. It is easy to see that the annihilator, level and the modified level of $\underline{L}$ are of the claimed form. Since $\underline{L}$ is even, $\omega\mathfrak{d}\mathfrak{c}^2$ is divisible by 2. Therefore, the last statement also holds true. □

Proposition 3.8 *Every homomorphism between* $\mathcal{O}$*-lattices of the form* (3.1) *is given by a multiplication of some nonzero element in* K*.*

Proof Let $(\mathfrak{c}, \omega)$ and $(\mathfrak{c}', \omega')$ be as in Definition 3.6. Let $\varphi : (\mathfrak{c}, \omega) \to (\mathfrak{c}', \omega')$ be a homomorphism. We can find a positive integer N such that $N\mathfrak{c}$ is integral. Since φ is an $\mathcal{O}$-module homomorphism, we have $N\varphi(x) = \varphi(Nx) = Nx\varphi(1)$ for all $x \in \mathfrak{c}$. Hence, $\varphi(x) = x\varphi(1)$ for all $x \in \mathfrak{c}$. In addition, since φ is isometric, we have $\omega xy = \omega'\varphi(x)\varphi(y) = \omega' xy\varphi(1)^2$ for all $x, y \in \mathfrak{c}$. This implies that for nonzero x and y, we have $\omega = \omega'\varphi(1)^2$. Since, ω and ω' are nonzero elements of K, we obtain that $\varphi(1) \neq 0$, i.e. φ is injective. □

Proposition 3.9 *Let* $(\mathfrak{c}, \omega)$ *and* $(\mathfrak{c}', \omega')$ *be as in Definition 3.6. Then the lattices* $(\mathfrak{c}, \omega)$ *and* $(\mathfrak{c}', \omega')$ *are isomorphic if and only if* $\omega' = a^2\omega$ *and* $\mathfrak{c}' = a^{-1}\mathfrak{c}$ *for some* $a \in K^*$*.*

Proof Suppose that φ is an isomorphism from $(\mathfrak{c}, \omega)$ to $(\mathfrak{c}', \omega')$. From Proposition 3.8, we have $\varphi(x) = xa$ $(x \in \mathfrak{c})$ for some nonzero $a \in K$. Since φ is a surjection, we have $\mathfrak{c}' = \mathfrak{c}a$. Moreover, since φ is isometric, we have $\omega xy =$

$\omega' \varphi(x)\varphi(y) = \omega' x y a^2$ for all $x, y \in \mathfrak{c}$. Then, by taking nonzero x and y, we have $\omega' = a^2\omega$. The other inclusion is obvious. □

Proposition 3.10 *Let $\underline{L} = (L, \beta)$ be a totally positive definite $\mathcal{O}$-lattice of rank one. Then $\underline{L}$ is isomorphic to an $\mathcal{O}$-lattice of the form $(\mathfrak{c}, \omega)$ as in Definition 3.6.*

Proof Since $\underline{L}$ has rank one, using [FT93, § II.4, Thm. 13(b)], we obtain that L is isomorphic to a fractional $\mathcal{O}$-ideal, say $\mathfrak{c}$. Let φ be an isomorphism from $\mathfrak{c}$ onto L. Note that there exists a positive integer N' such that $N'\mathfrak{c}$ is integral. Then for any $x \in \mathfrak{c}$, we have $N'\varphi(x) = \varphi(N'x) = N'x\varphi(1)$, i.e. $\varphi(x) = x\varphi(1)$. Note that since φ is an isomorphism, $a := \varphi(1) \neq 0$. Let $c \in \mathfrak{c}$ such that $\beta(ca, ca) \neq 0$, and let $x, y \in \mathfrak{c}$. We can find a positive integer N such that $Nx, Ny, Nc \in \mathcal{O}$. Then we have

$$\begin{aligned} Nc^2\beta(\varphi(x), \varphi(y)) = Nc^2\beta(ax, ay) = c\beta(Ncax, ay) = Nxc\beta(ca, ay) \\ = x\beta(ca, Ncay) = Nxy\beta(ca, ca). \end{aligned}$$

We set $\omega := \beta(ca, ca)/c^2$. Hence, φ defines an isomorphism from the lattice $\underline{L}$ onto $(\mathfrak{c}, \omega)$. The fact that $\omega\mathfrak{c}^2$ lies in $\mathfrak{d}^{-1}$ follows from the integrality of the $\mathcal{O}$-lattice $\underline{L}$, and that $\omega \gg 0$ follows from the totally positive definiteness of the $\mathcal{O}$-lattice $\underline{L}$. □

3.2 Algebraic Prerequisites

We set

$$\mathscr{C} := \mathbb{C} \otimes_{\mathbb{Q}} K, \quad \mathscr{R} := \mathbb{R} \otimes_{\mathbb{Q}} K.$$

We view $\mathscr{C}$ as a ring with respect to the multiplication induced from the map defined by $(z \otimes a, z' \otimes a') \mapsto (zz') \otimes (aa')$ $(z, z' \in \mathbb{C},\ a, a' \in K)$, and as algebra over $\mathbb{C}$ and over K via the maps satisfying $(z, z' \otimes a) \mapsto (zz') \otimes a$ and $(a, z \otimes a') \mapsto z \otimes (aa')$, respectively. In particular, we identify $\mathbb{C}$ and K with their images in $\mathscr{C}$ under the embeddings $z \mapsto z \otimes 1$ and $a \mapsto 1 \otimes a$. Similar conventions are made for $\mathscr{R}$, which we view as a subring of $\mathscr{C}$. In particular, the group $\mathcal{O}$ can be identified with its image in $\mathscr{C}$ under the embedding $b \mapsto 1 \otimes b$ $(b \in \mathcal{O})$. Hence the group Γ becomes a subgroup of $\mathrm{SL}(2, \mathscr{C})$ of 2×2-matrices over the ring $\mathscr{C}$ with determinant 1.

We use N and tr for the norm and trace of the $\mathbb{C}$-algebra $\mathscr{C}$. Thus, if c is an element of $\mathscr{C}$ and $f(x)$ is the characteristic polynomial of the endomorphism of $\mathscr{C}$ given by multiplication by c, then $f(x) = x^m - \mathrm{tr}(c)\, x^{m-1} + \cdots + (-1)^m \mathrm{N}(c)$. Likewise, if c is in $\mathscr{C}$, then

$$\mathrm{N}(c) = \prod_{\sigma \in \mathscr{E}} \sigma(c), \quad \mathrm{tr}(c) = \sum_{\sigma \in \mathscr{E}} \sigma(c).$$

Here $\mathscr{E}$ is the set of the $\mathbb{C}$-linear continuations of all embeddings $\sigma : K \hookrightarrow \mathbb{C}$ to $\mathbb{C}$-linear maps $\sigma : \mathscr{C} \to \mathbb{C}$ (we use the same letter for the embedding and its linear continuation). Thus $\sigma(z \otimes a) = z\sigma(a)$. The maps $\sigma \in \mathscr{E}$ are coordinate functions of the ring isomorphism $\prod_{\sigma\in\mathscr{E}} \sigma : \mathscr{C} \xrightarrow{\cong} \mathbb{C}^n$ ($n = [K : \mathbb{Q}]$), where we take coordinate wise multiplication as multiplication on $\mathbb{C}^n$. In particular, an element c of $\mathscr{C}$ is invertible (i.e. multiplication by c is an isomorphism of $\mathscr{C}$) if and only if $\mathrm{N}(c)$ is different from 0. The $\mathbb{Q}$-bilinear map $(z, a) \mapsto \overline{z} \otimes a$ induces a $\mathbb{Q}$-linear involution on $\mathscr{C}$ which we also indicate by placing a bar over the argument. We set

$$\mathscr{H} := \{z \in \mathscr{C} : \Im\big(\sigma(z)\big) > 0, \text{ for all } \sigma \in \mathscr{E}\}.$$

Note that $\mathscr{H}$ is an open subset of $\mathscr{C}$.

Proposition 3.11 *The group $\Gamma_{\mathscr{R}}$ acts on $\mathscr{H}$ via*

$$(A, \tau) \mapsto A\tau := (a\tau + b)(c\tau + d)^{-1}. \tag{3.2}$$

Proof First of all, for c and d in $\mathscr{R}$ and τ in $\mathscr{H}$ the element $c\tau + d$ is invertible and $A\tau$ is in $\mathscr{H}$. For proving this we use the easily proved identity

$$(A\tau - \overline{A\tau})(c\tau + d)(c\overline{\tau} + d) = \tau - \overline{\tau}.$$

The left hand side under σ ($\sigma \in \mathscr{E}$) has hence strictly positive imaginary part. This shows that $\mathrm{N}(c\tau + d) \neq 0$, hence that $c\tau + d$ is invertible. So, we have $\Im\big(\sigma(A\tau)\big) = \Im\big(\sigma(\tau)\big)/\sigma\big((c\tau + d)(c\overline{\tau} + d)\big) > 0$ ($\sigma \in \mathscr{E}$).

Next, it obviously holds true that $1\tau = \tau$. Let $A = \left(\begin{smallmatrix} a & b \\ c & d \end{smallmatrix}\right)$, $B = \left(\begin{smallmatrix} a' & b' \\ c' & d' \end{smallmatrix}\right)$ be in $\Gamma_{\mathscr{R}}$ and $\tau \in \mathscr{H}$. Then we have

$$B(A\tau) = \frac{a'\frac{a\tau+b}{c\tau+d} + b'}{c'\frac{a\tau+b}{c\tau+d} + d'} = \frac{(a'a + b'c)\tau + a'b + db'}{(c'a + cd')\tau + c'b + dd'} = (BA)\tau.$$

This proves the proposition. □

Under the identification of $\mathscr{C}$ with $\mathbb{C}^n$ from above the set $\mathscr{H}$ corresponds to the subset of vectors w in $\mathbb{C}^n$ whose components have all positive imaginary part, and the trace and the norm of w become the sum and the product of its components, respectively. For $w \in \mathscr{C}$, we use $\sqrt{w}$ for the element in $\mathscr{C}$ which corresponds to the element $\big(\sqrt{w_1}, \ldots, \sqrt{w_n}\big)$ in $\mathbb{C}^n$ under the above isomorphism which sends w to $\big(w_1, \ldots, w_n\big)$. For the choice of the square root of a complex number, we refer to the section "Notations".

For an $\mathscr{O}$-lattice $\underline{L} = (L, \beta)$ of rank r, the $\mathscr{O}$-module $L_{\mathscr{C}} := \mathscr{C} \otimes_{\mathscr{O}} L$ (similarly, $L_{\mathscr{R}} := \mathscr{R} \otimes_{\mathscr{O}} L$) becomes a $\mathscr{C}$-module via $\mathscr{C}$-linear continuation of the following map

$$(w', w \otimes x) \mapsto w'w \otimes x \quad (w, w' \in \mathscr{C}, x \in L),$$

which contains L and K as $\mathscr{O}$-submodules via the identifications $x \mapsto 1 \otimes x$ and $a \mapsto (1 \otimes a) \otimes 1$ $(x \in L, a \in K)$, respectively. Moreover, $L_{\mathscr{C}}$ becomes a $\mathbb{C}$-vector space of dimension nr via linear continuation of the following map

$$(z, w \otimes x) \mapsto zw \otimes x \quad (z \in \mathbb{C}, w \in \mathscr{C}, x \in L).$$

3.3 The Metaplectic Cover $\tilde{\Gamma}_{\mathscr{R}}$ of $\Gamma_{\mathscr{R}}$

Definition 3.12 The *metaplectic cover of* $\Gamma_{\mathscr{R}}$ (resp. Γ) is the set of tuples of the form $\left(A = \left(\begin{smallmatrix} a & b \\ c & d \end{smallmatrix}\right), w\right)$, where $A \in \Gamma_{\mathscr{R}}$ (resp. $A \in \Gamma$) and $w : \mathscr{H} \to \mathbb{C}$ a holomorphic function satisfying $w^2(\tau) = \mathrm{N}(c\tau + d)$, with the following operation

$$(A, w) \cdot (B, v) := \big(AB, w(B\tau)v(\tau)\big).$$

(Here $B\tau$ denotes the group action (3.2).) In the following we denote the metaplectic cover of $\Gamma_{\mathscr{R}}$ (resp. Γ) by $\tilde{\Gamma}_{\mathscr{R}}$ (resp. $\tilde{\Gamma}$).

Remark Note that $\Gamma_{\mathscr{R}}$ (resp. Γ) is in fact a group.

The group $\tilde{\Gamma}_{\mathscr{R}}$ is a central extension of $\Gamma_{\mathscr{R}}$

$$1 \to \langle(1, -1)\rangle \to \tilde{\Gamma}_{\mathscr{R}} \xrightarrow{\pi} \Gamma_{\mathscr{R}} \to 1, \tag{3.3}$$

where π is the map $(A, w) \mapsto A$ and the second arrow is the inclusion. In the following, for $A = \left(\begin{smallmatrix} a & b \\ c & d \end{smallmatrix}\right) \in \Gamma_{\mathscr{R}}$, we write $A^* := \left(A, \mathrm{N}\left(\sqrt{c\tau + d}\right)\right)$. Note that the map $A \mapsto A^*$ from $\Gamma_{\mathscr{R}}$ to $\tilde{\Gamma}_{\mathscr{R}}$ does not in general define a group homomorphism.

We know from Sect. 3.2 that Γ can be embedded into $\Gamma_{\mathscr{R}}$. Hence, the group $\tilde{\Gamma}$ can also be viewed as a subgroup of $\Gamma_{\mathscr{R}}$. For later use we determine a set of generators for $\tilde{\Gamma}$.

Proposition 3.13 *The group $\tilde{\Gamma}$ is generated by the elements $T_b^* = (T_b, 1)$ $(b \in \mathscr{O})$, $S^* = \left(S, \mathrm{N}\left(\sqrt{\tau}\right)\right)$ and $I := (1, -1)$. If the degree n of K over $\mathbb{Q}$ is odd, then $\tilde{\Gamma}$ is already generated by T_b^* and S^*.*

Proof Let $A = \left(\begin{smallmatrix} a & b \\ c & d \end{smallmatrix}\right) \in \Gamma$ and π be the projection in (3.3). We know from Theorem 2.1 that $A = \pi(A^*)$ can be written as a word in $S = \pi(S^*)$ and $T_b = \pi(T_b^*)$ $(b \in \mathscr{O})$. Hence, A^* can be written as a word in $(T_b)^*$, S^* and an element lying in the kernel of π, which equals $(1, \pm 1)$. Since every element of $\tilde{\Gamma}$ is either of the form A^*, or $A^* I$, the first statement holds true. If n is odd, then $I = (S^*)^4$, i.e. the second statement holds true. □

3.4 The Jacobi Group of an $\mathcal{O}$-Lattice

In the present section, we shall define the Heisenberg group and the Jacobi group associated to an $\mathcal{O}$-lattice. Moreover, we shall study various actions of these groups which are important in the sequel.

As explained in the section "Notations", we shall use $e\{c\}$ for $\exp(2\pi i \operatorname{tr}(c))$, where $c \in \mathcal{C}$.

Definition 3.14 Let $\underline{L} = (L, \beta)$ be an $\mathcal{O}$-lattice. The *Heisenberg group associated to* $\underline{L}$ is

$$H(L_{\mathcal{R}}) := \{(x, y, \xi) : x, y \in L_{\mathcal{R}}, \xi \in \mathbb{C}^*\}$$

together with the operation

$$(x, y, \xi) \cdot (x', y', \xi') = \left(x + x', y + y', \xi\xi' e\left\{(\beta(x, y') - \beta(x', y))/2\right\}\right). \tag{3.4}$$

Moreover, we use $H(L^\#)$ and $H(L)$ for the subgroups

$$H(L^\#) = \left\{(x, y, \xi) : x, y \in L^\#, \xi \in \mu_{2l}\right\}$$
$$H(L) = \left\{(x, y, e\{\beta(x, y)/2\}) : x, y \in L\right\}.$$

Here l is the exponent of the abelian group $L^\#/L$.

Remark The exact sequence

$$0 \to \mathbb{C}^* \to H(L_{\mathcal{R}}) \to L_{\mathcal{R}} \times L_{\mathcal{R}} \to 0$$
$$\xi \mapsto (x, y, \xi) \mapsto (x, y)$$

shows that $H(L_{\mathcal{R}})$ is a central extension of $L_{\mathcal{R}} \times L_{\mathcal{R}}$ by $\mathbb{C}^*$.

Note that the elements $(x, 0, 1)$ and $(0, y, 1)$ $(x, y \in L)$ generate $H(L)$. Note also that $H(L)$ is a normal subgroup of $H(L^\#)$.

Proposition 3.15 *The operation* (3.4) *defines indeed a group structure on* $H(L_{\mathcal{R}})$.

Proof The neutral element is $(0, 0, 1)$. For an element (x, y, ξ) of $H(L_{\mathcal{R}})$, the inverse element equals $(-x, -y, \xi^{-1})$. The associativity follows from the following identity:

$$\begin{aligned}
&(x, y, \xi) \cdot \left((x', y', \xi') \cdot (x'', y'', \xi'')\right) \\
&= \left(x + x' + x'', y + y' + y'', \xi\xi'\xi'' e\left\{(\beta(x', y'') - \beta(x'', y'))/2\right\} \times\right. \\
&\qquad \left. \times\, e\left\{(\beta(x, y' + y'') - \beta(x' + x'', y))/2\right\}\right) \\
&= \left(x + x' + x'', y + y' + y'', \xi\xi'\xi'' \times\right. \\
&\qquad \left. \times\, e\left\{(\beta(x + x', y'') + \beta(x, y') - \beta(x'', y + y') - \beta(x', y))/2\right\}\right)
\end{aligned}$$

$$= \left(x + x', y + y', \xi\xi' e\left\{(\beta(x, y') - \beta(x', y))/2\right\}\right) \cdot (x'', y'', \xi'')$$
$$= \left((x, y, \xi) \cdot (x', y', \xi')\right) \cdot (x'', y'', \xi'').$$

□

For later use we note the following

Proposition 3.16 *Let $\underline{L} = (L, \beta)$ be an $\mathcal{O}$-lattice and let l be the exponent of $L^\#/L$. Then $H(L^\#)/H(L)$ is a central extension of $L^\#/L$ by μ_{2l}. More precisely the applications $\xi \mapsto (0, \xi)H(L)$ and $(x, \xi)H(L) \mapsto x + L$ define an exact sequence*

$$1 \longrightarrow \mu_{2l} \longrightarrow H(L^\#)/H(L) \longrightarrow L^\#/L \longrightarrow 1.$$

The order of $H(L^\#)/H(L)$ equals, in particular, $2l|L^\#/L|^2$.

Remark It is not hard to show that, for even $\underline{L}$, the group $H(L^\#)/H(L)$ is also a central extension of the Heisenberg group $H(D_{\underline{L}})$ associated to the $\mathcal{O}$-FQM $D_{\underline{L}}$ (see Definition 2.61).

Proof of Proposition 3.16 The proposition follows easily by a straightforward calculation. □

Proposition 3.17 *Let $\underline{L} = (L, \beta)$ be an $\mathcal{O}$-lattice. The group $\Gamma_{\mathcal{R}}$ acts on the group $H(L_{\mathcal{R}})$ from the right via*

$$\left((x, y, \xi), A\right) \rightarrow (x, y, \xi)^A := \left((x, y)A, \xi\right).$$

Here, $(x, y)A$ stands for the formal multiplication of the row vector (x, y) and the matrix A. More precisely, for $A = \left(\begin{smallmatrix} a & b \\ c & d \end{smallmatrix}\right)$, we use $(x, y)A = (ax + cy, bx + dy)$.

Proof To prove the proposition we need to show that the axioms of a group action are satisfied, and that, for fixed $A \in \Gamma_{\mathcal{R}}$, the map $h \mapsto h^A$ defines a group homomorphism of $H(L_{\mathcal{R}})$. Let $A, B \in \Gamma_{\mathcal{R}}$ and $h \in H(L_{\mathcal{R}})$. Write $A = \left(\begin{smallmatrix} a & b \\ c & d \end{smallmatrix}\right)$, $B = \left(\begin{smallmatrix} a' & b' \\ c' & d' \end{smallmatrix}\right)$ and $h = (x, y, \xi)$. It is obvious that $h^1 = h$, and, since we have

$$\begin{aligned}((x, y)^A)^B &= ((xa + yc)a' + (xb + yd)c', (xa + yc)b' + (xb + yd)d') \\ &= ((aa' + bc')x + (ca' + dc')y, (ab' + bd')x + (cb' + dd')y) \\ &= (x, y)^{AB},\end{aligned}$$

the first part holds true.

Next we need to show $h^A \cdot h'^A = (h \cdot h')^A$ $(h' \in H(L_{\mathcal{R}}))$. Write $h' = (x', y', \xi')$. The identity

$$e\left\{(\beta(ax + cy, bx' + dy' - \beta(ax' + cy', bx + dy))/2\right\}$$
$$= e\left\{(\beta(x, y') - \beta(x', y))/2\right\}$$

clearly proves the second part. □

Definition 3.18 Let $\underline{L} = (L, \beta)$ be an $\mathcal{O}$-lattice. We use $J(L_{\mathbb{R}})$ for the semidirect product of $\Gamma_{\mathbb{R}}$ and $H(L_{\mathbb{R}})$ with respect to the action in Proposition 3.17, in short

$$J(L_{\mathbb{R}}) = \Gamma_{\mathbb{R}} \ltimes H(L_{\mathbb{R}}).$$

Similarly, we use $J(L^{\#}) := \Gamma \ltimes H(L^{\#})$ and, if $\underline{L}$ is even $J(L) := \Gamma \ltimes H(L)$.

Remark Recall from the general definition of semidirect products that $J(L_{\mathbb{R}})$ consists of all pairs (A, h) of elements A in $\Gamma_{\mathbb{R}}$ and h in $H(L_{\mathbb{R}})$ together with the operation

$$(A, h) \cdot (B, h') = (AB, h^B \cdot h').$$

More explicitly, the operation can be written as

$$\begin{aligned}\big(A, (x, y, \xi)\big) \cdot \big(B, (x', y', \xi')\big) = \big(&AB, ax + cy + x', bx + dy + y', \\ &\xi\xi' e\left\{\big(\beta(ax + cy, y') - \beta(x', bx + dy)\big)/2\right\}\big).\end{aligned}$$

For the definition of $J(L^{\#})$ and $J(L)$ to make sense, we need that $H(L^{\#})$ and $H(L)$ are invariant under the action of Γ on the Heisenberg group. For $H(L^{\#})$ this is always true, whereas for $H(L)$ this holds true only if $\underline{L}$ is even. Indeed, if $h = (x, y, e\{\beta(x, y)/2\})$ is in $H(L)$, then, for A in Γ, we have that h^A lies in $H(L)\,(0, e\{ab\beta(x) + cd\beta(y)\})$. But $e\{ab\beta(x) + cd\beta(y)\}$ equals 1 for all A in Γ and all x and y in L if and only if $\underline{L}$ is even.

Note that $J(L^{\#})$ and $J(L)$ are subgroups of $J(L_{\mathbb{R}})$. We view $\Gamma_{\mathbb{R}}$ and $H(L_{\mathbb{R}})$ as subgroups of $J(L_{\mathbb{R}})$ via the maps $A \mapsto (A, 1)$ and $h \mapsto (1, h)$, respectively. So, when we write Ah, we mean the element $(A, 1) \cdot (1, h)$.

Lemma 3.19 *The map $\gamma : \Gamma_{\mathbb{R}} \times \mathcal{H} \to \mathcal{H}$, defined by*

$$\gamma(A, \tau) := c\tau + d \quad (A = \left(\begin{smallmatrix} a & b \\ c & d \end{smallmatrix}\right))$$

satisfies the following identity (cocycle identity):

$$\gamma(A, B\tau)\gamma(B, \tau) = \gamma(AB, \tau). \tag{3.5}$$

Proof Let $A = \left(\begin{smallmatrix} a & b \\ c & d \end{smallmatrix}\right)$, $B = \left(\begin{smallmatrix} a' & b' \\ c' & d' \end{smallmatrix}\right)$ be elements of $\Gamma_{\mathbb{R}}$. The identity (3.5) holds true, since we have:

$$\begin{aligned}\gamma(A, B\tau)\gamma(B, \tau) &= (cB\tau + d)(c'\tau + d') \\ &= ca'\tau + cb' + dc'\tau + dd' = (ca' + dc')\tau + cb' + dd' \\ &= \gamma(AB, \tau).\end{aligned}$$

□

Proposition 3.20 *Let $\underline{L} = (L, \beta)$ be an $\mathcal{O}$-lattice. The group $\Gamma_{\mathcal{R}}$ acts on $\mathcal{H} \times L_{\mathcal{C}}$ via*

$$(A, (\tau, z)) \mapsto A(\tau, z) := \left(A\tau, \frac{z}{\gamma(A, \tau)}\right).$$

Moreover, $H(L_{\mathcal{R}})$ also acts on $\mathcal{H} \times L_{\mathcal{C}}$ via

$$((x, y, \xi), (\tau, z)) \mapsto (x, y, \xi)(\tau, z) := (\tau, z + x\tau + y).$$

Proof Let $A = \left(\begin{smallmatrix} a & b \\ c & d \end{smallmatrix}\right)$, $B = \left(\begin{smallmatrix} a' & b' \\ c' & d' \end{smallmatrix}\right)$ be elements of $\Gamma_{\mathcal{R}}$ and $(\tau, z) \in \mathcal{H} \times L_{\mathcal{C}}$. Since obviously we have $1(\tau, z) = (\tau, z)$, the following identity proves the first statement:

$$B(A(\tau, z)) = \left(B(A\tau), \frac{\frac{z}{\gamma(A,\tau)}}{\gamma(B, A\tau)}\right) = \left(BA\tau, \frac{z}{\gamma(BA, \tau)}\right) = BA(\tau, z).$$

Here to obtain the second identity we used (3.5). For proving the second statement, let $h = (x, y, \xi)$ and $h' = (x', y', \xi')$ be elements of $H(L_{\mathcal{R}})$. Obviously, we have $1(\tau, z) = (\tau, z)$. Furthermore, we calculate

$$\begin{aligned} h'(h(\tau, z)) &= (x', y', \xi')((x, y, \xi)(\tau, z)) = (\tau, z + (x + x')\tau + (y + y')) \\ &= (x' + x, y' + y, \xi'\xi e\left\{(\beta(x', y) - \beta(x, y'))/2\right\})(\tau, z) \\ &= (h' \cdot h)(\tau, z). \end{aligned}$$

This proves the proposition. □

Lemma 3.21 *Let $\underline{L} = (L, \beta)$ be an $\mathcal{O}$-lattice. For any $y \in \mathcal{H} \times L_{\mathcal{C}}$, $h \in H(L_{\mathcal{R}})$ and $A \in \Gamma_{\mathcal{R}}$, we have*

$$h(Ay) = A(h^A y).$$

Proof Write $y = (\tau, z)$, $h = (x, y, \xi)$ and $A = \left(\begin{smallmatrix} a & b \\ c & d \end{smallmatrix}\right)$. The claimed identity holds true, since we have

$$\begin{aligned} h(Ay) &= \left(A\tau, \frac{z}{\gamma(A, \tau)} + x(A\tau) + y\right) = \left(A\tau, \frac{z + (ax + cy)\tau + bx + dy}{\gamma(A, \tau)}\right) \\ &= A(\tau, z + (ax + cy)\tau + bx + dy) = A(h^A y). \end{aligned}$$

□

Proposition 3.22 *Let $\underline{L} = (L, \beta)$ be an $\mathcal{O}$-lattice. The group $J(L_{\mathcal{R}})$ acts on $\mathcal{H} \times L_{\mathcal{C}}$ via*

$$((A, h), (\tau, z)) \mapsto (A, h)(\tau, z) := A(h(\tau, z)).$$

Proof Let (A, h), (B, h') be in $J(L_{\mathscr{R}})$ and $u \in \mathscr{H} \times L_{\mathscr{C}}$. Since obviously we have $(1, 1)u = u$, the following identity proves the proposition

$$\begin{aligned}(B, h')\big((A, h)u\big) &= (B, h')\big(A(hu)\big) = B\big(h'\big(A(hu)\big)\big),\\ &= B(A(h'^{A}(hu))) = (BA)((h'^{A} \cdot h)(u))\\ &= (BA, h'^{A} \cdot h)(u) = \big((B, h') \cdot (A, h)\big)(u).\end{aligned}$$

Here we used Lemma 3.21 to obtain the third identity, and for the forth identity we used the second part of Proposition 3.20. □

Proposition 3.23 *Let* $k \in \mathbb{Z}$. *The group* $\Gamma_{\mathscr{R}}$ *acts from the right on the space* Hol$(\mathscr{H})$ *via*

$$(\phi, A) \mapsto (\phi|_k A)(\tau) := \mathrm{N}\left(\gamma(A, \tau)\right)^{-k} \phi(A\tau).$$

(See (3.19) *for the function* $\gamma(A, \tau)$*).*

Proof Let $A, B \in \Gamma_{\mathscr{R}}$, $\phi \in$ Hol$(\mathscr{H})$ and $\tau \in$ Hol$(\mathscr{H})$. The following identity proves the proposition, since we obviously have $\phi|_k 1 = \phi$:

$$\begin{aligned}\Big((\phi|_k A)|_k B\Big)(\tau) &= \mathrm{N}\left(\gamma(B, \tau)\right)^{-k} \mathrm{N}\left(\gamma(A, B\tau)\right)^{-k} \phi(A(B\tau))\\ &= \mathrm{N}\left(\gamma(AB, \tau)\right)^{-k} \phi(AB\tau) = (\phi|_k AB)(\tau).\end{aligned}$$

The second identity follows from (3.5) and Proposition 3.11. □

Proposition 3.24 *Let* $k \in \mathbb{Z}$ *and* $\underline{L} = (L, \beta)$ *be an* $\mathscr{O}$*-lattice. The group* $\Gamma_{\mathscr{R}}$ *acts from the right on* Hol$(\mathscr{H} \times L_{\mathscr{C}})$ *via*

$$(\phi, A) \mapsto (\phi|_{k,\underline{L}} A)(\tau, z) := \mathrm{N}\left(\gamma(A, \tau)\right)^{-k} e\left\{\frac{-c\beta(z)}{\gamma(A, \tau)}\right\} \phi\left(A\tau, \frac{z}{\gamma(A, \tau)}\right),$$

where $A = \left(\begin{smallmatrix} a & b \\ c & d \end{smallmatrix}\right)$. *(Recall that* $\beta(z) = \frac{1}{2}\beta(z, z)$*.)*

Proof Let $\phi \in$ Hol$(\mathscr{H} \times L_{\mathscr{C}})$ and $B \in \Gamma_{\mathscr{R}}$. Write $B = \left(\begin{smallmatrix} a' & b' \\ c' & d' \end{smallmatrix}\right)$. Since $\phi|_k 1 = \phi$, the following identity implies the proposition

$$\begin{aligned}&\Big((\phi|_{k,\underline{L}} A)|_{k,\underline{L}} B\Big)(\tau, z)\\ &= \mathrm{N}\left(\gamma(B, \tau)\right)^{-k} \mathrm{N}\left(\gamma(A, B\tau)\right)^{-k} e\left\{\frac{-c\beta(z)}{\gamma(B, \tau)^2 \gamma(A, B\tau)}\right\} e\left\{\frac{-c'\beta(z)}{\gamma(B, \tau)}\right\}\\ &\quad \times \phi\left(AB\tau, \frac{z}{\gamma(AB, \tau)}\right)\end{aligned}$$

$$= \mathrm{N}(\gamma(AB,\tau))^{-k} e\left\{\frac{-(ca'+dc')\beta(z)}{\gamma(AB,\tau)}\right\}\phi\Big(AB\tau, \frac{z}{\gamma(AB,\tau)}\Big)$$
$$= \big(\phi|_{k,\underline{L}}AB\big)(\tau,z).$$

To obtain the second identity we used (3.5) and Proposition 3.11, also the identity

$$\frac{ca'+dc'}{\gamma(AB,\tau)} = \frac{c'}{\gamma(B,\tau)} + \frac{c}{\gamma(A,B\tau)\gamma(B,\tau)^2},$$

where we used $a'd' - b'c' = 1$ and (3.5). □

Proposition 3.25 *Let $k \in \mathbb{Z}$ and let $\underline{L} = (L,\beta)$ be an $\mathscr{O}$-lattice. The group $H(L_{\mathscr{R}})$ acts from the right on* $\mathrm{Hol}(\mathscr{H}\times L_{\mathscr{C}})$ *via*

$$\big(\phi,(x,y,\xi)\big) \mapsto \big(\phi|_{k,\underline{L}}(x,y,\xi)\big)(\tau,z)$$
$$:= \xi e\left\{\tau\beta(x)+\beta(x,z)+\beta(x,y)/2\right\}\phi(\tau,z+x\tau+y).$$

Proof Let $\phi \in \mathrm{Hol}(\mathscr{H}\times L_{\mathscr{C}})$ and $h,h' \in H(L_{\mathscr{R}})$. Write $h = (x,y,\xi)$ and $h' = (x',y',\xi')$. Since we obviously have $\phi|_{k,\underline{L}}1 = \phi$, the following identity proves that we have indeed an action:

$$\begin{aligned}\Big(\big(\phi|_{k,\underline{L}}h\big)|_{k,\underline{L}}h'\Big)(\tau,z) &= \xi\xi' e\left\{\tau\beta(x+x')+\beta(x+x',z)+\beta(x,y')\right\}\times\\ &\quad\times e\left\{\beta(x',y')/2+\beta(x,y)/2\right\}\times\\ &\quad\times\phi(\tau,z+(x+x')\tau+y+y')\\ &= \xi\xi' e\left\{\big(\beta(x,y')-\beta(x',y)\big)/2\right\}\times\\ &\quad\times e\left\{\tau\beta(x+x')+\beta(x+x',z)+\beta(x+x',y+y')/2\right\}\\ &\quad\times\phi(\tau,z+(x+x')\tau+y+y')\\ &= \big(\phi|_{k,\underline{L}}(h\cdot h')\big)(\tau,z).\end{aligned}$$

□

Lemma 3.26 *Let $k \in \mathbb{Z}$ and let $\underline{L} = (L,\beta)$ be an $\mathscr{O}$-lattice. For any element ϕ in* $\mathrm{Hol}(\mathscr{H}\times L_{\mathscr{C}})$, *$A \in \Gamma_{\mathscr{R}}$, $h \in H(L_{\mathscr{R}})$, we have*

$$\big(\phi|_{k,\underline{L}}h\big)|_{k,\underline{L}}A = \big(\phi|_{k,\underline{L}}A\big)|_{k,\underline{L}}h^A.$$

Proof Let $y \in \mathscr{H}\times L_{\mathscr{C}}$. Write $y = (\tau,z)$, $h = (x,y,\xi)$ and $A = \left(\begin{smallmatrix}a & b\\ c & d\end{smallmatrix}\right)$. We have on the left

$$\Big((\phi|_{k,\underline{L}}h)|_{k,\underline{L}}A\Big)(\tau,z) = \xi\,\mathrm{N}\left(\gamma(A,\tau)\right)^{-k} e\left\{\frac{-c\beta(z)}{\gamma(A,\tau)}\right\} \times$$
$$e\left\{A\tau\beta(x) + \beta\left(x, \frac{z}{\gamma(A,\tau)}\right) + \beta(x,y)/2\right\}\phi\left(A\tau, \frac{z}{\gamma(A,\tau)} + xA\tau + y\right).$$

Since $h^A = (xa + yc, xb + yd, \xi)$, on the right we have

$$\begin{aligned}\Big((\phi|_{k,\underline{L}}A)|_{k,\underline{L}}h^A\Big)(\tau,z)&\\ = \xi e\left\{\tau\beta(xa+yc) + \beta(xa+yc,z) + \beta(xa+yc,xb+yd)/2\right\}&\times\\ \mathrm{N}\left(\gamma(A,\tau)\right)^{-k} e\left\{\frac{-c\beta(z+(xa+yc)\tau+xb+yd)}{\gamma(A,\tau)}\right\}&\times\\ \times\,\phi\left(A\tau, \frac{z+(xa+yc)\tau+xb+yd}{\gamma(A,\tau)}\right)&.\end{aligned}$$

The claimed identity follows now from the following identities:

$$\frac{z}{\gamma(A,\tau)} + xA\tau + y = \frac{z+(xa+yc)\tau+xb+yd}{\gamma(A,\tau)},$$
$$-c\beta(z+(xa+yc)\tau+xb+yd) + \tau(c\tau+d)\beta(xa+yc) +$$
$$(c\tau+d)\beta(xa+yc,z) + \frac{c\tau+d}{2}\beta(xa+yc,xb+yd)$$
$$= -c\beta(z) + (a\tau+b)\beta(x) + \beta(x,z) + \frac{c\tau+d}{2}\beta(x,y).$$

The first one is obvious. The second one follows using $ad - bc = 1$. □

Proposition 3.27 *Let $k \in \mathbb{Z}$ and let $\underline{L} = (L,\beta)$ be an $\mathcal{O}$-lattice. The group $J(L_{\mathcal{R}})$ acts from the right on* $\mathrm{Hol}(\mathcal{H} \times L_{\mathcal{C}})$ *via*

$$\big(\phi,(A,h)\big) \mapsto \phi|_{k,\underline{L}}(A,h) := \big(\phi|_{k,\underline{L}}A\big)|_{k,\underline{L}}h.$$

Proof Let $\phi \in \mathrm{Hol}(\mathcal{H} \times L_{\mathcal{C}})$, $(B,h') \in J(L_{\mathcal{R}})$. From Propositions 3.24 and 3.25, we have $\phi|_{k,\underline{L}}(1,1) = \phi$. Moreover, we have (writing $|$ for $|_{k,\underline{L}}$)

$$\begin{aligned}\big(\phi|(A,h)\big)|(B,h') &= \Big(\big((\phi|A)|h\big)|B\Big)|h' = \Big(\big((\phi|A)|B\big)|h^B\Big)|h'\\ &= (\phi|AB)|(h^Bh') = \phi|(AB,h^Bh') = \phi|\big((A,h)(B,h')\big).\end{aligned}$$

The first and fourth identities follow from the very definition of the $J(L_{\mathcal{R}})$-action. For the second identity we used Lemma 3.26, for the third identity we used Propositions 3.24 and 3.25. □

If we replace the integer k in Proposition 3.23 with a half integer, then the action does not anymore define an action because of the ambiguity of the square root of $\mathrm{N}(c\tau + d)$. To solve the problem of this square root, we have to pass to the metaplectic cover $\tilde{\Gamma}_{\mathcal{R}}$ of $\Gamma_{\mathcal{R}}$ (recall Sect. 3.3 for its definition). For a number k in $\frac{1}{2}\mathbb{Z}$, we define the action $\big((A,w),\phi\big) \mapsto \phi|_k((A,w)$ of $\tilde{\Gamma}_{\mathcal{R}}$ on $\mathrm{Hol}(\mathcal{H})$ and $\mathrm{Hol}(\mathcal{H} \times L_{\mathcal{C}})$ as in the Propositions 3.23 and 3.24, respectively, but with the factor $\mathrm{N}\left(c\tau + d\right)^{-k}$ replaced by $w(\tau)^{-2k}$. It is clear that this defines indeed an action. Thus we can state

Proposition 3.28 *Let $\underline{L} = (L,\beta)$ be an $\mathcal{O}$-lattice and $k \in \frac{1}{2}\mathbb{Z}$. The group $\tilde{\Gamma}_{\mathcal{R}}$ acts on the right of the space* $\mathrm{Hol}(\mathcal{H} \times L_{\mathcal{C}})$ *via*

$$\big(\phi,(A,w)\big) \mapsto \phi|_{k,\underline{L}}(A,w)(\tau,z)$$
$$:= w(\tau)^{-2k} e\left\{\frac{-c\beta(z)}{\gamma(A,\tau)}\right\} \phi\left(A\tau, \frac{z}{\gamma(A,\tau)}\right). \tag{3.6}$$

Definition 3.29 Let $\underline{L} = (L,\beta)$ be an $\mathcal{O}$-lattice. The semidirect product of $\tilde{\Gamma}_{\mathcal{R}}$ and $H(L_{\mathcal{R}})$ with respect to the action

$$\big((x,y,\xi),(A,w)\big) \to (x,y,\xi)^{(A,w)} := \big((x,y)A,\xi\big) \tag{3.7}$$

is denoted by $\tilde{J}(L_{\mathcal{R}})$, and is called the *Jacobi group associated to* $\underline{L}$. We set also $\tilde{J}(L^{\#}) := \tilde{\Gamma} \ltimes H(L^{\#})$ and, if $\underline{L}$ is even, $\tilde{J}(L) := \tilde{\Gamma} \ltimes H(L)$.

Remark We view $\tilde{\Gamma}_{\mathcal{R}}$ and $H(L_{\mathcal{R}})$ as subgroups of $\tilde{J}(L_{\mathcal{R}})$ via the maps $\alpha \mapsto \big(\alpha, 1\big)$ and $h \mapsto (1,h)$, respectively. So, when we write αh, we mean the element (α,h).

Remark If we combine the action in (3.6) with the action of $H(L_{\mathcal{R}})$ on the space $\mathrm{Hol}(\mathcal{H} \times L_{\mathcal{C}})$ (see Proposition 3.25), we obtain a right action of $\tilde{J}(L_{\mathcal{R}})$ on $\mathrm{Hol}(\mathcal{H} \times L_{\mathcal{C}})$.

We shall now define certain differential operators on the space of smooth complex valued functions $\mathbb{C}^{\infty}(\mathcal{H} \times L_{\mathcal{C}})$. For this end let $\mathcal{E}$ denote the set of all $\mathbb{C}$-linear extensions of the embeddings from K into $\mathbb{R}$ to $\mathbb{C}$-linear maps from $\mathcal{C}$ into $\mathbb{C}$ (as already introduced in Sect. 3.2). Note that $\mathrm{tr} \circ \beta : L_{\mathcal{C}} \times L_{\mathcal{C}} \to \mathbb{C}$ is the $\mathbb{C}$-bilinear continuation of the non-degenerate $\mathbb{Z}$-linear form $(x,y) \mapsto \mathrm{tr} \circ \beta(x,y)$ from $L \times L \to \mathbb{Z}$ (see Proposition 3.2) we conclude that $\mathrm{tr} \circ \beta$ is non-degenerate on $L_{\mathcal{C}} \times L_{\mathcal{C}}$. It is then easy to prove that there is a basis of the $\mathbb{C}$-vector space $L_{\mathcal{C}}$ with coordinate functions $z_{\sigma,j}$ ($j = 1,\ldots,r$, $\sigma \in \mathcal{E}$) such that, for any σ in $\mathcal{E}$ and all z_1 and z_2 in $L_{\mathcal{C}}$, we have

$$\sigma \circ \beta(z_1,z_2) = \sum_{j=1}^{r} z_{\sigma,j}(z_1) z_{\sigma,j}(z_2). \tag{3.8}$$

We view $z_{\sigma,j}$ also as functions on $\mathscr{H} \times L_{\mathscr{C}}$ by setting $z_{\sigma,j}(\tau, z) = z_{\sigma,j}(z)$ for τ in $\mathscr{H}$ and $z \in L_{\mathscr{C}}$. Furthermore, we use τ_σ for the function on $\mathscr{H} \times L_{\mathscr{C}}$ such that $\tau_\sigma(\tau, z) = \sigma(\tau)$. Note that

$$\{\tau_\sigma\}_{\sigma\in\mathscr{E}} \times \{z_{\sigma,j}\}_{\sigma\in\mathscr{E},1\le j\le r} : \mathscr{H} \times L_{\mathscr{C}} \to \mathbb{H}^n \times \mathbb{C}^{nr}$$

defines a biholomorphic map. Here $\mathbb{H}$ denotes the usual upper half plane in $\mathbb{C}$ and r and n are the rank of L and the degree of K over $\mathbb{Q}$, respectively. For σ in $\mathscr{E}$ we set

$$\Delta_\sigma := \frac{\partial^2}{\partial z_{\sigma,1}^2} + \ldots + \frac{\partial^2}{\partial z_{\sigma,r}^2}, \tag{3.9}$$

$$H_\sigma := \frac{\partial}{\partial \tau_\sigma} - \frac{1}{4\pi i}\Delta_\sigma, \tag{3.10}$$

and call these operators the σ-*Laplace operator* and σ-*Heat operator* on $\mathscr{H} \times L_{\mathscr{C}}$, respectively.

Lemma 3.30 *For $\sigma \in \mathscr{E}$, $\tau \in \mathscr{H}$ and s in $L_{\mathscr{C}}$, we have the following formulas:*

$$H_\sigma\, e\{\tau\beta(s) + \beta(s,z)\} = 0 \tag{3.11}$$

$$H_\sigma\, e\{-\beta(z)/\tau\} = \frac{r}{2\tau_\sigma} e\{-\beta(z)/\tau\}. \tag{3.12}$$

Here the expressions on the left after the differential operators are considered as functions in (τ, z) on $\mathscr{H} \times L_{\mathscr{C}}$.

Proof As immediate consequences of (3.8) and $\operatorname{tr}\left(\beta(z)/\tau\right) = \sum_{\sigma\in\mathscr{E}} \frac{1}{2\tau_\sigma} \sum_{j=1}^{r} z_{\sigma,j}^2$ one obtains

$$\Delta_\sigma\, e\{\beta(s,z)\} = 2(2\pi i)^2\, \sigma\big(\beta(s)\big)\, e\{\beta(s,z)\},$$

$$\frac{\partial}{\partial \tau_\sigma} e\{\tau\beta(s)\} = 2\pi i\, \sigma\big(\beta(s)\big)\, e\{\tau\beta(s)\},$$

$$\Delta_\sigma\, e\{-\beta(z)/\tau\} = -\frac{2\pi i}{\tau_\sigma}\left(r - \frac{4\pi i\, \sigma\beta(z)}{\tau_\sigma}\right) e\{-\beta(z)/\tau\},$$

$$\frac{\partial}{\partial \tau_\sigma} e\{-\beta(z)/\tau\} = \frac{2\pi i\, \sigma\beta(z)}{\tau_\sigma^2} e\{-\beta(z)/\tau\}.$$

The claimed identities of the lemma are now obvious. □

Proposition 3.31 *Let $\underline{L} = (L, \beta)$ be an $\mathscr{O}$-lattice of rank r, and let σ in $\mathscr{E}$. Then for any $\phi \in \mathrm{Hol}(\mathscr{H} \times L_{\mathscr{C}})$, and $\alpha = (A, w) \in \tilde{\Gamma}$, we have*

$$H_\sigma(\phi|_{r/2,\underline{L}}\alpha) = \sigma\big(\gamma(A,\tau)\big)^{-2}(H_\sigma\phi)|_{r/2,\underline{L}}\alpha.$$

Proof It suffices to prove the claimed identity for the standard generators T_b^* (b in $\mathcal{O}$), I and S^* of $\tilde{\Gamma}$. Except for S^* the claimed identity is then obvious. For proving the identity for S^* we write first of all

$$\big(\phi|_{r/2,\underline{L}}S^*\big)(\tau,z) = \phi(-1/\tau, z/\tau)\, e\{-\beta(z)/\tau\}\, \mathrm{N}\big(\sqrt{\tau}\big)^{-r}.$$

Thus $\phi|_{r/2,\underline{L}}S^*$ is a product of three functions, which we denote by f_1, f_2 and f_3, respectively. Applying now the heat operator H_σ, yields accordingly

$$\begin{aligned} &H_\sigma\big(\phi|_{k,\underline{L}}S^*\big)(\tau,z)\\ &\quad = (H_\sigma f_1)\, f_2\, f_3 + f_1\,(H_\sigma f_2)\, f_3 - \frac{1}{2\pi i}(\nabla_\sigma f_1)\cdot(\nabla_\sigma f_2)\, f_3 + f_1\, f_2 \frac{\partial}{\partial \tau_\sigma} f_3, \end{aligned} \tag{3.13}$$

where $\nabla_\sigma = \big\{\frac{\partial}{\partial z_{\sigma,j}}\big\}_j$. A short calculation, using $E_\sigma = \sum_j z_{\sigma,j}\frac{\partial}{\partial z_{\sigma,j}}$, shows

$$\begin{aligned} H_\sigma f_1 &= \tau_\sigma^{-2}\big(H_\sigma\phi\big)(-1/\tau, z/\tau) + \tau_\sigma^{-2}\big(E_\sigma\phi\big)(-1/\tau, z/\tau),\\ H_\sigma f_2 &= \frac{r}{2\tau_\sigma} f_2,\\ -\frac{1}{2\pi i}(\nabla_\sigma f_1)\cdot(\nabla_\sigma f_2) &= \frac{-1}{\tau_\sigma^2}(E_\sigma\phi)(-1/\tau, z/\tau)\, f_2,\\ \frac{\partial}{\partial \tau_\sigma} f_3 &= -\frac{r}{2\tau_\sigma} f_3. \end{aligned}$$

Here, the second identity is identical with (3.12). We observe that the second and fourth term in (3.13) cancel. Moreover, the third term and the second term in the formula for $H_\sigma f_1$ multiplied by $f_2 f_3$ cancel. Finally, the remaining first term of $H_\sigma f_1$ multiplied by $f_2 f_3$ equals $\tau_\sigma^{-2}\big(H_\sigma\phi\big)|_{r/2,\underline{L}}S^*$. This proves the proposition. □

3.5 The Jacobi Theta Functions

In this section we introduce and study certain spaces of Jacobi theta functions which will be important in all remaining chapters. We shall show that these spaces of Jacobi theta functions are $\tilde{\Gamma}$-modules (see Theorem 3.1), and we shall calculate explicitly the matrix coefficients of the associated representations.

For $t \in \mathcal{C}$, we shall use q^t for the function on $\mathcal{H}$ such that

$$q^t(\tau) = e\{t\tau\}.$$

Definition 3.32 Let $\underline{L} = (L, \beta)$ be a totally positive definite even $\mathscr{O}$-lattice. For $x \in L^\#/L$, we set

$$\vartheta_{\underline{L},x}(\tau, z) := \sum_{\substack{s \in L^\# \\ s \equiv x \bmod L}} q^{\beta(s)} e\{\beta(s, z)\} \quad (\tau \in \mathscr{H}, z \in L_{\mathscr{C}}). \tag{3.14}$$

We refer to these functions as the *Jacobi theta functions associated to* $\underline{L}$. Moreover, we set

$$\Theta_{\underline{L}} := \operatorname{span}_{\mathbb{C}} \{\vartheta_{\underline{L},x} : x \in L^\#/L\}.$$

Remark It is easily verified that the series defining $\vartheta_{\underline{L},x}$ are absolutely convergent and that the $\vartheta_{\underline{L},x}$ are holomorphic (here one needs that $\underline{L}$ is totally positive definite). Note also that $\vartheta_{\underline{L},x}$ depends only on the residue class of x modulo L.

Proposition 3.33 *For fixed* τ, *the functions* $z \mapsto \vartheta_{\underline{L},x}(\tau, z)$ $(x \in L^\#/L)$ *defined in* (3.14) *are linearly independent. In particular, the dimension of the* $\mathbb{C}$*-vector space* $\Theta_{\underline{L}}$ *equals* $|L^\#/L|$.

Proof Fix τ in $\mathscr{H}$, and let ϕ_x $(x \in L^\#/L)$ be complex numbers. Set $\phi(z) := \sum_{x \in L^\#/L} \phi_x \vartheta_{\underline{L},x}(\tau, z)$. It is immediate from the definition of $\vartheta_{\underline{L},x}$ that, for $y \in L^\#$, we have $\vartheta_{\underline{L},x}(\tau, z + y) = \vartheta_{\underline{L},x}(\tau, z) e\{\beta(y, x)\}$. For each $x_0 \in L^\#$, we therefore have

$$\begin{aligned}
\sum_{y \in L^\#/L} \phi(z + y) e\{-\beta(y, x_0)\} &= \sum_{y \in L^\#/L} \sum_{x \in L^\#/L} \phi_x \vartheta_{\underline{L},x}(\tau, z + y) e\{-\beta(y, x_0\} \\
&= \sum_{y \in L^\#/L} \sum_{x \in L^\#/L} \phi_x \vartheta_{\underline{L},x}(\tau, z) e\{\beta(y, x - x_0\} \\
&= \phi_{x_0} |L^\#/L| \vartheta_{\underline{L},x_0}(\tau, z).
\end{aligned}$$

For the last identity we used Proposition 1.11. Hence, if $\phi(z)$ vanishes identically, then $\phi_{x_0} = 0$ unless $\vartheta_{\underline{L},x_0}(\tau, z)$ vanishes for all z. But the latter is impossible since $\vartheta_{\underline{L},x_0}(\tau, z)$, considered as a Fourier development in z, would vanish identically only if all coefficients $q^{\beta(r)}$ were identically zero. □

The main results of this section is the following theorem.

Theorem 3.1 *Let* $\underline{L} = (L, \beta)$ *be a totally positive definite even* $\mathscr{O}$*-lattice of rank* r. *The space* $\Theta_{\underline{L}}$ *is a* $\tilde{\Gamma}$*-module. More precisely, for* x *in* $L^\#$ *and* α *in* $\tilde{\Gamma}$, *say,* α *equals* $\left(\left(\begin{smallmatrix} a & b \\ c & d \end{smallmatrix}\right), \epsilon \operatorname{N}\left(\sqrt{c\tau + d}\right)\right)$, *we have*

$$\begin{aligned}
\vartheta_{\underline{L},x}|_{r/2,\underline{L}}\alpha = c_{z_\alpha}^{\underline{L}}(\alpha) \sum_{y \in L^\#/L} e\{(\beta(ax + cy, bx + dy) - \beta(x, y))/2\} \times \\
\times e\{\beta(bx + dy, z_\alpha)\} \vartheta_{\underline{L}, z_\alpha + ax + cy},
\end{aligned}$$

where

$$c^{\underline{L}}_{z_\alpha}(\alpha) = \frac{e\{-bd\beta(z_\alpha)\}}{|S_c/L|} \lim_{t\to\infty} \left(\vartheta_{\underline{L},-dz_\alpha}|\alpha\right)(it \otimes 1, 0).$$

For z_α and S_c, we refer to Lemma 3.41 below.

Before we give the proof of this theorem at the end of the section we deduce various consequences.

Corollary 3.34 *Let $n = [K : \mathbb{Q}]$. We have*

(i) $\vartheta_{\underline{L},x}|_{r/2,\underline{L}} T_b^* = e\{b\beta(x)\}\, \vartheta_{\underline{L},x} \quad (b \in \mathscr{O})$
(ii) $\vartheta_{\underline{L},x}|_{r/2,\underline{L}} S^* = \frac{1}{\sqrt{|L^\#/L|}} i^{-nr/2} \sum_{y\in L^\#/L} e\{-\beta(y,x)\}\, \vartheta_{\underline{L},y}$
(iii) $\vartheta_{\underline{L},x}|_{r/2,\underline{L}} I = (-1)^r \vartheta_{\underline{L},x}$.

Proof The formulas in (i), (iii) are immediate consequence of Theorem 3.1. For (ii), note that the element z_α can be taken to be zero (since here $S_c = L$; see Lemma 3.41). Comparing the formula of the theorem for $\alpha = S^*$ and (ii) shows that it remains to prove

$$\lim_t \vartheta_{\underline{L},0}|_{r/2,\underline{L}} S^*(it \otimes 1, 0) = \frac{1}{\sqrt{|L^\#/L|}} i^{-nr/2}.$$

However, this is an immediate consequence of the transformation formula [Ebe02, Prop. 5.7]. □

Corollary 3.35 *We carry over the notations of Theorem 3.1. Let $\mathfrak{l}$ denote the level of $\underline{L}$, and let $\tilde{\Gamma}_0(\mathfrak{l})$ be the inverse image of $\Gamma_0(\mathfrak{l}) := \left(\begin{smallmatrix} \mathscr{O} & \mathscr{O} \\ \mathfrak{l} & \mathscr{O} \end{smallmatrix}\right) \cap \mathrm{SL}(2, \mathscr{O})$ in $\tilde{\Gamma}$.*

(i) There exists a linear character ε of $\Gamma_0(\mathfrak{l})$ such that

$$\vartheta_{\underline{L},x}|_{r/2,\underline{L}}\alpha = \varepsilon(\alpha) e\{ab\beta(x)\}\, \vartheta_{\underline{L},ax}$$

for all α in $\tilde{\Gamma}_0(\mathfrak{l})$.

(ii) There exists a quadratic Dirichlet character χ modulo $\mathfrak{l}$ such that, for $\alpha = (\left(\begin{smallmatrix} a & b \\ c & d \end{smallmatrix}\right), w)$, one has $\varepsilon(\alpha)^q = \chi(d)$, where $q = 2$ if the rank r of $\underline{L}$ is odd, and $q = 1$ otherwise.

(iii) Set

$$\Gamma_{\underline{L}} = \{\alpha = (A, w) \in \tilde{\Gamma} : A \in \Gamma(\mathfrak{l}),\ \varepsilon(\alpha) = 1\}.$$

The projection of $\Gamma_{\underline{L}}$ on $\Gamma(\mathfrak{l})$ is surjective. The group $\Gamma_{\underline{L}}$ acts trivially on $\Theta_{\underline{L}}$.

Remark Note that $\Gamma_{\underline{L}}$ is normal. Indeed, $\Gamma_{\underline{L}}$ equals the group of all α is the inverse image of $\Gamma(\mathfrak{l})$ in $\tilde{\Gamma}$ which fix $\Theta_{\underline{L}}$ point wise.

Proof Since c is in L we can choose $z_\alpha = 0$. Using that, for all y in $L^\#$, we have cy in L and $c\beta(y)$ in $\mathfrak{d}^{-1}$ (since c is in $\mathfrak{l}$), the transformation formula of the theorem simplifies to $\vartheta_{\underline{L},x}|_{r/2,\underline{L}}\alpha = \varepsilon(\alpha)(e\{ab\beta(x)\}\,\vartheta_{\underline{L},ax}$, where $\varepsilon(\alpha) = c_0^{\underline{L}}(\alpha)|L^\#/L|$. It is clear that $\varepsilon(\alpha)$ is a linear character of $\Gamma_0(\mathfrak{l})$.

By [Ebe02, Prop. 5.10] we know that $\varepsilon(\alpha)^q = \chi(d)$ for a quadratic Dirichlet character modulo $\mathfrak{l}$ (for deducing this from Prop. 5.10 in [Ebe02] let $\underline{L}_2 := (L \oplus L, \beta_2)$, where $\beta_2(x, y) = \beta(x) + \beta(y)$, if r is odd, and let $\underline{L}_2 = \underline{L}$ otherwise, and choose loc. cit. $V = K \otimes (L \oplus L)$ and $\Gamma = L \oplus L$; note that [Ebe02] only treats lattices of even rank which holds true for $\underline{L}_2$).

For proving (iii) we note that ε is trivial on the inverse image $\tilde{\Gamma}(\mathfrak{l})$ of $\Gamma(\mathfrak{l})$ if r is even. Otherwise ε is quadratic on $\tilde{\Gamma}(\mathfrak{l})$, and $\varepsilon(I) = -1$. But then, for A in $\Gamma(\mathfrak{l})$ we have $\varepsilon(A^*) = 1$ or $\varepsilon(A^*I) = 1$. This proves the corollary. □

For the proof of Theorem 3.1 we shall need some preparations. One of the main tools is the action of the Heisenberg group $H(L^\#)$ on $\Theta_{\underline{L}}$, which we shall now explain.

Proposition 3.36 *The application* $(\vartheta, (x, y, \xi)) \mapsto \vartheta|_{r/2,\underline{L}}(x, y, \xi)$ *(where r is the rank of $\underline{L}$) defines a right $H(L^\#)$-module structure on the space $\Theta_{\underline{L}}$. More precisely, we have*

$$\vartheta_{\underline{L},x'}|_{r/2,\underline{L}}(x, y, \xi) = \xi e\left\{\beta(x, y)/2 + \beta(x', y)\right\}\vartheta_{\underline{L},x'+x}. \tag{3.15}$$

The group $H(L)$ acts, in particular, trivially on $\Theta_{\underline{L}}$.

Remark The space $\Theta_{\underline{L}}$ can thus be viewed as an $H(L^\#)/H(L)$-module. Recall that $H(L^\#)/H(L)$ is a finite group of order $2l|L^\#/L|^2$ where l is the exponent of $L^\#/L$ (see Proposition 3.16) .

Proof of Proposition 3.36 Using the very definition of the $|_{r/2,\underline{L}}$-action of the Heisenberg group (see Proposition 3.25) we find

$$\begin{aligned}
&\vartheta_{\underline{L},x'}|_{r/2,\underline{L}}(x, y, \xi)(\tau, z)\\
&= \xi e\{\tau\beta(x) + \beta(x, z) + \beta(x, y)/2\} \sum_{\substack{s\in L^\# \\ s\equiv x' \bmod L}} q^{\beta(s)} e\{\beta(s, z + x\tau + y)\}\\
&= \xi e\{\beta(x, y)/2\} \sum_{\substack{s\in L^\# \\ s\equiv x' \bmod L}} q^{\beta(s+x)} e\{\beta(s + x, z)\}\, e\{\beta(s, y)\}\\
&= \xi e\{\beta(x, y)/2\}\, e\left\{\beta(x', y)\right\}\vartheta_{\underline{L},x+x'}(\tau, z).
\end{aligned}$$

The last identity follows on noting that, for $s \equiv x' \bmod L$, we have that $e\{\beta(s, y)\}$ equals $e\{\beta(x', y)\}$, and by substituting $s - x$ for s in the sum. This proves the formula for the action of $H(L^\#)$.

Recall that $H(L)$ is generated by the elements $\left(x, y, e\left\{\frac{1}{2}\beta(x,y)\right\}\right)$ $(x, y \in L)$. But for these elements h and by the just proved formulas we obviously have $\vartheta_{\underline{L},x'}|_{r/2,\underline{L}}h = \vartheta_{\underline{L},x'}$. This proves the second statement. □

Proposition 3.37 *The character $\chi_{H(L^{\#})}$ of the $H(L^{\#})$-module $\Theta_{\underline{L}}$ satisfies*

$$\chi_{H(L^{\#})}(x, y, \xi) = \begin{cases} \xi\,|L^{\#}/L|\,e\{\beta(x,y)/2\} & \text{if } x, y \in L \\ 0 & \text{otherwise.} \end{cases}$$

In particular, $\Theta_{\underline{L}}$ is an irreducible $H(L^{\#})$-right module.

Remark Note that the formula for $\chi_{H(L^{\#})}$ implies that, for any α in $\tilde{\Gamma}$ and h in $H(L^{\#})$, we have $\chi_{H(L^{\#})}(\alpha^{-1}h\alpha) = \chi_{H(L^{\#})}(h)$.

Proof of Proposition 3.37 By the formula (3.15) for the action of $H(L^{\#})$ we have

$$\operatorname{tr}\left((x, y, \xi), \Theta_{\underline{L}}\right) = \begin{cases} \xi e\{\beta(x,y)/2\} \sum_{x' \in L^{\#}/L} e\{\beta(x', y)\} & \text{if } x \in L \\ 0 & \text{otherwise.} \end{cases}$$

But the sum in the above identity is zero unless $y \in L$ (see Proposition 1.11) which proves the first statement.

For the second statement it suffices to prove that $\Theta_{\underline{L}}$, viewed as module over the finite group $H(L^{\#})/H(L)$ is irreducible. Indeed, we have

$$\frac{1}{|H(L^{\#})/H(L)|} \sum_{h \in H(L^{\#})/H(L)} |\operatorname{tr}\left(h, \Theta_{\underline{L}}\right)|^2 = \frac{|L^{\#}/L|^2}{|H(L^{\#})/H(L)|} \sum_{\xi \in \mu_{2l}} 1 = 1,$$

which implies the irreducibility [FH91, Cor. 2.15]. □

Lemma 3.38 *Let U denote the subgroup $0 \times L^{\#} \times 1$ of $H(L^{\#})$. For any α in $\tilde{\Gamma}$, the space $\Theta_{\underline{L}}^{\alpha^{-1}U\alpha}$ of functions in $\Theta_{\underline{L}}$ which are invariant under the subgroup $\alpha^{-1}U\alpha$ of $H(L^{\#})$ is one dimensional.*

Proof As already in the proof of the preceding proposition we view $\Theta_{\underline{L}}$ as module over the finite group $G := H(L^{\#})/H(L)$. Let π be the canonical projection from $H(L^{\#})$ onto G. We then have $\Theta_{\underline{L}}^{V} = \Theta_{\underline{L}}^{\pi(V)}$, where $V = \alpha^{-1}\pi(U)\alpha$. But then, by standard representation theory (see Corollary 2.20), we have

$$\dim \Theta_{\underline{L}}^{\pi(V)} = \frac{1}{|\pi(U)|} \sum_{v \in \pi(V)} \operatorname{tr}\left(v, \Theta_{\underline{L}}\right) = \frac{1}{|\pi(U)|} \sum_{u \in \pi(U)} \operatorname{tr}\left(u, \Theta_{\underline{L}}\right).$$

The second identity is an immediate consequence of the invariance under conjugation with α of the character of $H(L^{\#})$ as explained in the remark after

Proposition 3.37. But by the same proposition tr $(u, \Theta_{\underline{L}}) = |\pi(U)|$ if u is the neutral element of G, and tr $(u, \Theta_{\underline{L}}) = 1$ otherwise. The lemma is now obvious. □

Recall from the previous section that $\mathscr{E}$ denotes the set of the $\mathbb{C}$-linear extensions to $\mathscr{C} = \mathbb{C} \otimes_{\mathbb{Q}} K$ of the ($\mathbb{Q}$-linear) embeddings of K into the field of real numbers. Recall also that, for each σ in $\mathscr{E}$ we have associated the Heat operator H_σ [see (3.10)].

Lemma 3.39 *For any $\sigma \in \mathscr{E}$, and for any ϑ in $\Theta_{\underline{L}}$, we have*

$$H_\sigma \vartheta = 0.$$

Proof Since every ϑ in $\Theta_{\underline{L}}$ has a Fourier development in terms of the functions $e\{\tau\beta(s) + \beta(s,z)\}$ ($s \in L^\#$), and these functions are annihilated by H_σ (see Lemma 3.30) the lemma is obvious. □

Proposition 3.40 *Let ϕ be a holomorphic function on $\mathscr{H} \times L_{\mathscr{C}}$. Then following statements are equivalent:*

(i) $\phi \in \Theta_{\underline{L}}$
(ii) There is a half integral k such that $\phi|_{k,\underline{L}} h = \phi$ for all $h \in H(L)$, and $H_\sigma \phi = 0$ for all $\sigma \in \mathscr{E}$.

Proof Recall that $\Theta_{\underline{L}}$ has basis $\vartheta_{\underline{L},x}$ ($x \in L^\#/L$).

(i) $\Longrightarrow$ (ii). The invariance property follows from Proposition 3.36, which states that $H(L)$ acts trivially on $\Theta_{\underline{L}}$ with $k = r/2$. The second property is the preceding lemma.

(ii) $\Longrightarrow$ (i). Since ϕ is fixed under the action of $H(L)$, we have, in particular, $\phi(\tau,z) = \phi|_{k,\underline{L}}(0,y,1)(\tau,z) = \phi(\tau, z+y)$ for any $y \in L$. Hence, we can write

$$\phi(\tau,z) = \sum_{s\in L^\#} \phi_s(\tau) q^{\beta(s)} e\{\beta(s,z)\}$$

for suitable functions $\phi_s(\tau)$ on $\mathscr{H}$. By the same assumption again, for any $x \in L$, we have $\phi(\tau,z) = \phi|_{k,\underline{L}}(x,0,1)(\tau,z) = \phi(\tau, z + x\tau) e\{\tau\beta(x) + \beta(x,z)\}$. But this implies

$$\begin{aligned}\phi(\tau,z) &= \sum_{s\in L^\#} \phi_s(\tau) q^{\beta(s)} e\{\beta(s, z + x\tau)\}\, e\{\tau\beta(x) + \beta(x,z)\}\\ &= \sum_{s\in L^\#} \phi_s(\tau) q^{\beta(s+x)} e\{\beta(s+x, z)\}.\end{aligned}$$

Since this implies that the functions ϕ_s depend only on s modulo L, we then have

$$\phi(\tau,z) = \sum_{s\in L^\#/L} \phi_s(\tau) \vartheta_{\underline{L},s}(\tau,z).$$

The functions ϕ_s are holomorphic functions on $\mathscr{H}$. Indeed, $\phi(\tau, z)$ is holomorphic and we have, for any $s \in L^\#/L$,

$$\phi_s(\tau) = \frac{\sum_{y \in L^\#/L} \phi(\tau, z + y) e\{-\beta(y, s)\}}{|L^\#/L|\, \vartheta_{\underline{L},s}(\tau, z)}$$

(see the proof of Proposition 3.33). Now by the second assumption and Lemma 3.39, we obtain

$$0 = H_\sigma \phi(\tau, z) = \sum_{s \in L^\#/L} \left(\frac{\partial}{\partial \tau_\sigma} \phi_s(\tau) \right) \vartheta_{\underline{L},s}(\tau, z).$$

Since the $\vartheta_{\underline{L},s}(\tau, z)$, for fixed τ as functions of z, are linearly independent (Proposition 3.33), we deduce that the ϕ_s are constants, and hence that ϕ lies in $\Theta_{\underline{L}}$. □

The characterization of $\Theta_{\underline{L}}$ as given in the preceding proposition enables us to prove now that $\Theta_{\underline{L}}$ is invariant under $\tilde{\Gamma}$.

Proof of Theorem 3.1 We show that $\Theta_{\underline{L}}$ is invariant under the $|_{r/2,\underline{L}}$ action of $\tilde{\Gamma}$. Let ϑ in $\Theta_{\underline{L}}$ and $\alpha \in \tilde{\Gamma}$, and set $\phi := \vartheta|_{r/2,\underline{L}}\alpha$. We have to show that ϕ is an element of $\Theta_{\underline{L}}$. By Proposition 3.40, it suffices to show that ϕ is invariant under the action of $H(L)$, and that, for any σ in $\mathscr{E}$, we have $H_\sigma \phi = 0$. Let $h \in H(L)$. The first claim holds true, since we have (writing $|$ for $|_{r/2,\underline{L}}$)

$$\phi|h = (\vartheta|\alpha)|h = (\vartheta|\alpha h \alpha^{-1})|\alpha = \vartheta|\alpha = \phi.$$

The third identity follows from the fact that $\tilde{\Gamma}$ leaves $H(L)$ invariant under conjugation, and the fact that $H(L)$ acts trivially on $\Theta_{\underline{L}}$ (see Proposition 3.36).

The second claim also holds true, since we have

$$H_\sigma(\phi) = H_\sigma(\vartheta|\alpha) = \sigma\big(\gamma(A, \tau)\big)^{-2} (H_\sigma \vartheta)|\alpha = 0.$$

Here the second identity follows from Proposition 3.31, and the last one follows from Lemma 3.39. This proves the first part of the theorem. For deducing the explicit formulas for the action of $\tilde{\Gamma}$ we need some further preparations. □

Lemma 3.41 *For every* $\alpha = \left(\left(\begin{smallmatrix} a & b \\ c & d \end{smallmatrix}\right), w\right)$, *there exists* $z_\alpha \in L^\#$ *satisfying the congruence* $\operatorname{tr}\big(cd\beta(y)\big) \equiv \operatorname{tr}\big(d\beta(y, z_\alpha)\big) \bmod \mathbb{Z}$ *for all* y *in* $S_c := \{y \in L^\# : cy \in L\}$.

Proof The map $\varphi : S_c/L \to \frac{1}{2}\mathbb{Z}/\mathbb{Z}$, $y + L \mapsto \operatorname{tr}\big(cd\beta(y)\big) + \mathbb{Z}$ is a group homomorphism. Indeed, for $y, y' \in S_c$, we have

$$\varphi(y + y' + L) = \operatorname{tr}\big(cd\beta(y) + cd\beta(y') + cd\beta(y, y')\big) + \mathbb{Z},$$

and $cy \in L$ implies then $c\beta(y, y') \in \mathfrak{d}^{-1}$. We can continue φ to a group homomorphism $\tilde{\varphi} : L^\#/L \to \mathbb{Q}/\mathbb{Z}$ [Ser73, Ch. VI,§ 1, Prop. 1]. Since the tr $(D_{\underline{L}})$ is non-degenerate (see Proposition 3.2) the map $L^\#/L \to \mathrm{Hom}(L^\#/L, \mathbb{Q}/\mathbb{Z})$, $y + L \mapsto \beta(y, \_) + \mathbb{Z}$ is injective. Since a finite abelian group and its dual have the same order (see [Ser73, Ch. VI,§ 1, Prop. 2]), this map is an isomorphism. Hence there exists a z in $L^\#$ such that $\tilde{\varphi}(y) = \beta(y, z) + \mathbb{Z}$. Set $z_\alpha = az$. Then, for y in S_c, we have tr $(\beta(y, z)) \equiv$ tr $(d\beta(y, z_\alpha))$ mod $\mathbb{Z}$ since $ad \equiv 1 \bmod c$. □

Proof of Theorem 3.1 (cont.) It remains to calculate the matrix coefficients of the $\tilde{\Gamma}$-action on $\Theta_{\underline{L}}$. We prove first of all, that

$$\vartheta_{\underline{L},0}|\alpha = c_{z_\alpha}^{\underline{L}}(\alpha) \sum_{y \in L^\#/L} \vartheta_{\underline{L},z_\alpha}|\alpha^{-1}(0, y, 1)\alpha, \tag{3.16}$$

where $c_{z_\alpha}^{\underline{L}}(\alpha)$ is a constant, and z_α is as in Lemma 3.41. Here and in the following we write $|$ for $|_{r/2,\underline{L}}$.

For the proof denote the sum on the right hand side by S. Note that each term depends indeed only on the coset of y in $L^\#/L$ as follows easily from the invariance of $\vartheta_{\underline{L},z_\alpha}$ under $H(L)$.

The claimed identity follows from the fact that both sides are invariant under the subgroup $\alpha^{-1}(0 \times L^\# \times 1)\alpha$ of $H(L^\#)$, and that the space of functions in $\Theta_{\underline{L}}$ invariant under this subgroup is one dimensional (cf. Lemma 3.38). The invariance of the left hand side follows from the fact that $\vartheta_{\underline{L},0}$ is invariant under $0 \times L^\# \times 1$ (as follows from Proposition 3.36). The invariance of the right hand side follows from Proposition 2.15.

For concluding the proof of the formula we still have to show that S is different from zero. Writing $\alpha^{-1}(0, y, 1)\alpha = (cy, dy, 1)$, we obtain

$$\begin{aligned} S &= \sum_{y \in L^\#/L} \vartheta_{\underline{L},z_\alpha}|(cy, dy, 1) = \sum_{y \in L^\#/L} e\{\beta(cy, dy)/2\}\, e\{\beta(dy, z_\alpha)\}\, \vartheta_{\underline{L},z_\alpha + cy} \\ &= \sum_{x \in L^\#/L} \vartheta_{\underline{L},x} \sum_{\substack{y \in L^\#/L \\ z_\alpha + cy \equiv x \bmod L}} e\{cd\beta(y) + d\beta(y, z_\alpha)\}. \end{aligned}$$

From this we see that $S \neq 0$ since the $\vartheta_{\underline{L},x}$ are linearly independent, and since the inner sum is different from zero for $x = z_\alpha$. Indeed, in this case the inner sum runs over a complete set of representatives for S_c/L, and then, by the very definition of z_α, the terms are all equal to 1 (since tr $(cd\beta(y) + d\beta(y, z_\alpha)) \equiv$ tr $(2cd\beta(y)) \equiv 0 \bmod \mathbb{Z}$).

Next, note that, for any $x \in L^\#$, one has $\vartheta_{\underline{L},x} = \vartheta_{\underline{L},0}|(x, 0, 1)$. Using this identity and (3.16), we obtain

$$\begin{aligned}\vartheta_{\underline{L},x}|\alpha &= \vartheta_{\underline{L},0}|\alpha|\big(\alpha^{-1}(x,0,1)\alpha\big)\\ &= c_{z_\alpha}^{\underline{L}}(\alpha)\sum_{y\in L^\#/L}\vartheta_{\underline{L},z_\alpha}|\big(\alpha^{-1}(0,y,1)\alpha\big)|\big(\alpha^{-1}(x,0,1)\alpha\big)\\ &= c_{z_\alpha}^{\underline{L}}(\alpha)\sum_{y\in L^\#/L}\vartheta_{\underline{L},z_\alpha}|\big(ax+cy,bx+dy,e\{-\beta(x,y)/2\}\big).\end{aligned}$$

Applying again the formulas (3.15) for the $H(L^\#)$-action on $\Theta_{\underline{L}}$ we obtain thus

$$\begin{aligned}\vartheta_{\underline{L},x}|\alpha = c_{z_\alpha}^{\underline{L}}(\alpha)\sum_{y\in L^\#/L} e\{\beta(bx+dy,z_\alpha)\}\times\\ e\big\{\big(\beta(ax+cy,bx+dy)-\beta(x,y)\big)/2\big\}\,\vartheta_{\underline{L},z_\alpha+ax+cy},\end{aligned}\tag{3.17}$$

which is the formula stated in the theorem. Hence, it remains to calculate the constant $c_{z_\alpha}^{\underline{L}}(\alpha)$.

For obtaining a formula for $c_{z_\alpha}^{\underline{L}}(\alpha)$ we set $x=-dz_\alpha$ in (3.17), evaluate the resulting identity at $z=0$ and $\tau=it\otimes 1$ with real t, and let t tend to infinity. For calculating the limit of the right hand side of (3.17) we note that $\vartheta_{\underline{L},z_\alpha-adz_\alpha+cy}=\vartheta_{\underline{L},c(-bz_\alpha+y)}$, and that

$$\begin{aligned}&\lim_{t\mapsto\infty}\vartheta_{\underline{L},c(-bz_\alpha+y)}(it\otimes 1,0)\\ &\qquad= \lim_{t\mapsto\infty}\sum_{s\equiv c(-bz_\alpha+y)\bmod L} e^{-2\pi t\,\mathrm{tr}\left(\beta(s)\right)} = \begin{cases}1 & \text{if } c(-bz_\alpha+y)\in L\\ 0 & \text{otherwise.}\end{cases}\end{aligned}$$

But $c(-bz_\alpha+y)\in L$ if and only if $y\in bz_\alpha+S_c$. Note that $(ax+cy,bx+dy)=(-z_\alpha+ct,dt)$ for $(x,y)=(-dz_\alpha,bz_\alpha+t)$. Thus, the limit of the right hand side of (3.17) (specialized to $(\tau,z)=(it\otimes 1,0)$ and $x=-dz_\alpha$) becomes

$$\begin{aligned}c_{z_\alpha}^{\underline{L}}(\alpha)\sum_{t\in L^\#/L,ct\in L} e\{\beta(dt,z_\alpha)\}\,e\big\{\big(\beta(-z_\alpha+ct,dt)-\beta(-dz_\alpha,bz_\alpha+t)\big)/2\big\}\\ = c_{z_\alpha}^{\underline{L}}(\alpha)e\{bd\beta(z_\alpha)\}\,|S_c/L|.\end{aligned}$$

Summarizing we have found

$$c_{z_\alpha}^{\underline{L}}(\alpha)=\frac{e\{-bd\beta(z_\alpha)\}}{|S_c/L|}\lim_{t\to\infty}\big(\vartheta_{\underline{L},-dz_\alpha}|\alpha\big)(it\otimes 1,0),$$

which completes the proof of the theorem. □

We conclude this section by some propositions which we shall need in the last section of this chapter when we shall discuss the relation of Jacobi forms and vector-valued Hilbert modular forms.

Proposition 3.42 *The application*

$$\big(\vartheta_{\underline{L},x}, (\alpha, h)\big) \mapsto \vartheta_{\underline{L},x}|_{r/2,\underline{L}}(\alpha, h) := \big(\vartheta_{\underline{L},x}|_{r/2,\underline{L}}\alpha\big)|_{r/2,\underline{L}}h$$

defines a right $\tilde{J}(L^\#)$-module structure on $\Theta_{\underline{L}}$.

Proof This can be verified by a straightforward calculation similar to the one in the proof of Proposition 3.27 on using the Proposition 3.36 and the Theorem 3.1. □

Definition 3.43 By $\langle\cdot,\cdot\rangle$ we denote the Hermitian scalar product on $\Theta_{\underline{L}}$ which is anti-linear in the second argument, and which satisfies:

$$\langle\vartheta_{\underline{L},x}, \vartheta_{\underline{L},y}\rangle = \begin{cases} 1 & \text{if } x = y \\ 0 & \textit{otherwise}. \end{cases} \tag{3.18}$$

Proposition 3.44 *The $\tilde{J}(L^\#)$-action on $\Theta_{\underline{L}}$ is unitary with respect to the scalar product in* (3.18).

Proof For proving the invariance of the scalar product under the action of $\tilde{\Gamma}$ it suffice to prove the invariance under the generators T_b^*, I and S^* of $\tilde{\Gamma}$. For the generators T_b^* and I the invariance is obvious. For proving the invariance under S^*, let ϑ and ϑ' be elements of $\Theta_{\underline{L}}$, say, $\vartheta = \sum_{x\in L^\#/L} c(x)\vartheta_{\underline{L},x}$ and $\vartheta = \sum_{x'\in L^\#/L} c(x')\vartheta_{\underline{L},x'}$. Using the formula for the S^*-action from Corollary 3.34, we have

$$\vartheta|_{r/2,\underline{L}}S^* = \frac{i^{-nr/2}}{\sqrt{|L^\#/L|}} \sum_{x\in L^\#/L} c(x) \sum_{y\in L^\#/L} e\{-\beta(y,x)\}\,\vartheta_{\underline{L},y},$$

and similarly for ϑ'. Using these formulas we can write

$$\big\langle\vartheta|_{r/2,\underline{L}}S^*, \vartheta'|_{r/2,\underline{L}}S^*\big\rangle = \frac{1}{|L^\#/L|} \sum_{x,x'\in L^\#/L} c(x)\overline{c(x')} \sum_{y\in L^\#/L} e\left\{\beta(y, x'-x)\right\}.$$

By Proposition 1.11, the inner sum equals zero unless $x' = x$, when it equals $|L^\#/L|$. The right hand side becomes thus $\sum_{x,\in L^\#/L} c(x)\overline{c(x)}$, which equals indeed $\langle\vartheta, \vartheta'\rangle$. The invariance under $H(L^\#)$ can be easily deduced using the formulas for the action on $\Theta_{\underline{L}}$ from Proposition 3.36. □

3.6 Definition and Basic Properties of Jacobi Forms

In the present section we give finally the definition of Jacobi forms over totally real number fields, and we shall discuss their Fourier developments and theta expansions.

Definition 3.45 Let $\underline{L} = (L, \beta)$ be a totally positive definite even $\mathcal{O}$-lattice and let $k \in \frac{1}{2}\mathbb{Z}$. Moreover, let Δ be a subgroup of finite index in $\tilde{\Gamma}$, and let χ be a linear character of Δ whose kernel is of finite index in $\tilde{\Gamma}$. A *Jacobi form over K of weight k, index $\underline{L}$ and character χ on Δ* is a holomorphic function $\phi : \mathcal{H} \times L_{\mathcal{C}} \to \mathbb{C}$ satisfying

(i) $\big(\phi|_{k,\underline{L}}\alpha\big)(\tau, z) = \chi(\alpha)\phi(\tau, z) \quad (\alpha \in \Delta)$
(ii) $\big(\phi|_{k,\underline{L}}h\big)(\tau, z) = \phi(\tau, z) \quad (h \in H(L))$.

If $K = \mathbb{Q}$, we assume furthermore that the function ϕ is holomorphic at all cusps (see [EZ85]).

The $\mathbb{C}$-vector space of all Jacobi forms over K is denoted by $J^K_{k,\underline{L}}(\Delta, \chi)$.

(For the notion of $\mathcal{O}$-lattices we refer to Sect. 3.1, and for the space $L_{\mathcal{C}}$ we refer to Sect. 3.2. Moreover, for the actions of $\tilde{\Gamma}$ and $H(L)$ on the space $\mathrm{Hol}(\mathcal{H} \times L_{\mathcal{C}})$ we refer the reader to Propositions 3.28 and 3.25, respectively.)

If $\Delta = \tilde{\Gamma}$, we simply write $J^K_{k,\underline{L}}(\chi)$ for $J^K_{k,\underline{L}}(\Delta, \chi)$, and call this space the *space of Jacobi forms over K of weight k, index $\underline{L}$ and character χ*. In the following we shall mainly concentrate on the spaces $J^K_{k,\underline{L}}(\chi)$. If the number field in question is clear from the context, we refer to the Jacobi forms over K simply as Jacobi forms, and we write $J_{k,\underline{L}}(\Delta, \chi)$ instead of $J^K_{k,\underline{L}}(\Delta, \chi)$.

Remark Applying the transformation law (i) to $\alpha = (1, -1)$, we obtain, for ϕ in $J_{k,\underline{L}}(\chi)$, that $\chi(\alpha)\phi = \phi|_{k,\underline{L}}\alpha = (-1)^{2k}\phi$. Hence $J_{k,\underline{L}}(\chi)$ is trivial unless we have $\chi\big((1, -1)\big) = (-1)^{2k}$. If k is integral and $\chi\big((1, -1)\big) = (-1)^{2k}(= +1)$, then χ factors through a linear character $\underline{\chi}$ of Γ. In this case we can rewrite the transformation law (i) as $\phi|_{k,\underline{L}}A = \underline{\chi}(A)\phi \ (A \in \Gamma)$, and we shall also write $J_{k,\underline{L}}(\underline{\chi})$ for $J_{k,\underline{L}}(\chi)$. If k is not integral and $\chi\big((1, -1)\big) = (-1)^{2k}(= -1)$, then χ does not factor through a linear character of Γ (see Proposition 2.4).

Proposition 3.46 *Every Jacobi form ϕ in $J_{k,\underline{L}}(\chi)$ possesses a Fourier development of the form*

$$\phi(\tau, z) = \sum_{\substack{s \in L^{\#} \\ t \in h + \mathfrak{d}^{-1}}} c(t, s)\, q^t e\,\{\beta(s, z)\}. \tag{3.19}$$

Here h is an element of K such that $\chi(T_b) = e\,\{hb\}$ for all $b \in \mathcal{O}$.

Proof Set $\psi(\tau, z) = e\,\{-h\tau\}\,\phi(\tau, z)$. From the transformation laws in Definition 3.45, we have that $\psi(\tau, z)$ is periodic in τ and z with respect to $\mathcal{O}$ and L, respectively. Since $\psi(\tau, z)$ is holomorphic, it can be written as infinite sum of the functions $e\{t\tau + \beta(z, s)\}$, where t and s run through $\mathfrak{d}^{-1}$ and $L^{\#}$, respectively. □

Theorem 3.2 (Köcher Principle for Jacobi Forms) *Assume $K \neq \mathbb{Q}$. In the Fourier expansion* (3.19) *one has $c(t, s) = 0$ unless $t - \beta(s) \gg 0$ or $t = \beta(s)$.*

The proof of this theorem will be given in the next section, since it requires some extra tools which we have to develop first.

Remark For $K = \mathbb{Q}$, the statement $c(t, s) = 0$ unless $t - \beta(s) \gg 0$ or $t = \beta(s)$ is a part of the definition.

If ϕ is a Jacobi form of weight k and index $\underline{L}$ on a subgroup Δ of finite index in $\tilde{\Gamma}$, then ϕ possesses also a Fourier development of the form (3.19), where, however, the range of t will in general be different. The Köcher principle holds then also in this case. The proof for these statements is the same as in the case of Jacobi forms on the full Hilbert modular group, and is left to the reader. Furthermore, it is easy to see that, for γ in $\tilde{G}$, where $G = \mathrm{SL}(2, K)$, the function $\phi|_{k,\underline{L}}\gamma$ is again a Jacobi form of the same weight k and index $\underline{L}$ but on $\gamma^{-1}\Delta\gamma \cap \Delta$ (which has also finite index in $\tilde{\Gamma}$). In particular, $\phi|_{k,\underline{L}}\gamma$ has a Fourier expansion for which the Köcher principle holds.

Definition 3.47 Let ϕ be in $J_{k,\underline{L}}(\Delta, \chi)$. If, for each γ in $\widetilde{\mathrm{SL}(2, K)}$, the function $\phi|_{k,\underline{L}}\gamma$ has a Fourier development $\sum c(t, s)q^s e\{\beta(s, z)\}$ such that

$$c(t, s) = 0 \text{ unless } t - \beta(s) \gg 0,$$

then ϕ is called a *Jacobi cusp form*.

If ϕ has a Fourier development $\sum c(t, s)q^s e\{\beta(s, z)\}$ satisfying

$$c(t, s) = 0 \text{ unless } t = \beta(s),$$

then ϕ is called a *singular Jacobi form*.

Remark The vanishing condition for $\phi|_{k,\underline{L}}\gamma$ in the definition depends only on the double coset of the first component of γ in $\Delta_*\backslash\tilde{G}/\tilde{D}$, where Δ_* is the projection onto Γ and $G = \mathrm{SL}(2, K)$ as before, and D is the subgroup of triangular matrices in $\mathrm{SL}(2, K)$. To count these double cosets we identify G/D with the projective line $\mathbb{P}^1(K)$ via the application $\left(\begin{smallmatrix} a & b \\ c & d \end{smallmatrix}\right) \mapsto [a : c]$, so that the double coset space becomes the orbit space $\Delta_*\backslash\mathbb{P}^1(K)$ with respect to the natural action of G on the projective line. Note that there are only finitely many orbits since Δ_* has finite index in Γ. Indeed, $\Gamma\backslash\mathbb{P}^1(K)$ is in one to one correspondence with the ideal classes of K via the application which maps $[a : b]$ to the ideal class of $\mathcal{O}a + \mathcal{O}b$ (see e.g. [vdG88, Chap. I, Prop. 1.1]). In particular, if K has class number one and ϕ is a Jacobi form on the full modular group Γ, then ϕ is a cusp form if and only if its "singular part"

$$\phi_{\text{sing}} := \sum_{\substack{s \in L^{\#} \\ \beta(s) \in \mathfrak{d}^{-1}}} c(\beta(s), s)\, q^s e\{\beta(s, z)\}$$

vanishes identically.

Example 3.48 Let $\underline{L}$ be a totally positive definite even $\mathcal{O}$-lattice of rank r. For all $x \in L^{\#}/L$, the Jacobi theta functions $\vartheta_{\underline{L},x}$ associated to $\underline{L}$ as defined in (3.14) are singular Jacobi forms on the subgroup $\Gamma_{\underline{L}}$ of $\tilde{\Gamma}$ (see Corollary 3.35) of weight $r/2$. The invariance under the $\tilde{\Gamma}$-action follows from Corollary 3.35. The fact that they are singular and of weight $r/2$ is immediate from their very definition.

Theorem 3.3 *Let $\underline{L}$ be a totally positive definite even $\mathcal{O}$-lattice and ϕ in $J_{k,\underline{L}}(\chi)$. Then ϕ can be written in the form*

$$\phi(\tau, z) = \sum_{x \in L^{\#}/L} h_x(\tau) \vartheta_{\underline{L},x}(\tau, z), \tag{3.20}$$

where

$$h_x(\tau) = \sum_{d \in \beta(x) - h + \mathfrak{d}^{-1}} c\big(\beta(x) - d, x\big) q^{-d}$$

[with $c(t,s)$ as in (3.19)*]. The function h_x depends only on x modulo L.*

In the following we call the expansion (3.20) the *theta expansion of* ϕ.

Remark Note that the theorem implies that the Fourier coefficient $c(t, s)$ of a Jacobi form ϕ depends only on $\beta(s) - t$ and the coset $s + L$.

Proof of Theorem 3.3 Writing $d = \beta(s) - t$ and setting $C(d, s) := c\big(\beta(s) - d, s\big)$, we can write the Fourier development (3.19) in the form

$$\begin{aligned} \phi(\tau, z) &= \sum_{\substack{s \in L^{\#} \\ d \in \beta(s) - h + \mathfrak{d}^{-1}}} C(d, s)\, q^{\beta(s) - d} e\{\beta(s, z)\} \\ &= \sum_{x \in L^{\#}/L} \sum_{\substack{s \in L^{\#} \\ s \equiv x \bmod L}} q^{\beta(s)} e\{\beta(s, z)\} \sum_{d \in \beta(s) - h + \mathfrak{d}^{-1}} C(d, s)\, q^{-d}. \end{aligned} \tag{3.21}$$

Using the second transformation law in Definition 3.45 for elements $(x, 0, 1)$ ($x \in L$), we obtain

$$e\{\tau\beta(x) + \beta(x, z)\}\, \phi(\tau, z + x\tau) = \phi(\tau, z).$$

Inserting the Fourier development of ϕ into the left hand side, we obtain

$$e\{\tau\beta(x)+\beta(x,z)\} \sum_{\substack{s\in L^\# \\ d\in\beta(s)-h+\mathfrak{d}^{-1}}} C(d,s)\, q^{\beta(s)-d} e\{\beta(s,z+x\tau)\}$$

$$= \sum_{\substack{s\in L^\# \\ d\in\beta(s)-h+\mathfrak{d}^{-1}}} C(d,s)\, q^{\beta(s+x)-d} e\{\beta(s+x,z)\}\,.$$

Replacing s by $s-x$ and comparing the Fourier coefficients we obtain

$$C(d,s) = C(d,s-x) \quad (x \in L).$$

In other words, $C(d,s)$ depends only on s mod L. Thus the inner sum in (3.21) depends only on s mod L and equals hence h_x. But then (3.21) reads

$$\phi(\tau,z) = \sum_{x\in L^\#/L} h_x(\tau) \sum_{\substack{s\in L^\# \\ s\equiv x \bmod L}} q^{\beta(s)} e\{\beta(s,z)\}\,.$$

This proves the theorem. □

3.7 Jacobi Forms as Vector-Valued Hilbert Modular Forms

In the present section our main aim will be to set up an isomorphism between spaces of Jacobi forms and spaces of vector-valued Hilbert modular forms. In particular, this will imply the Köcher principle for Jacobi forms and that the spaces of Jacobi forms are finite dimensional. For explicit formulas for the dimensions of the spaces of Jacobi forms, the reader is referred to [SS14].

In the sequel we shall make use of various facts and notions concerning representations of groups which were recalled in Sect. 2.1.

Recall from Theorem 3.1 that the space $\Theta_{\underline{L}}$ spanned by the functions $\vartheta_{\underline{L},x}$ is invariant under $\tilde{\Gamma}$ with respect to the $|_{r/2,\underline{L}}$-action. Thus, for any α in $\tilde{\Gamma}$, there are numbers $\omega_{x,y}(\alpha)$ such that

$$\vartheta_{\underline{L},x}|_{r/2,\underline{L}}\alpha = \sum_{y\in L^\#/L} \omega(\alpha)_{x,y}\vartheta_{\underline{L},y} \qquad (x\in L^\#/L). \tag{3.22}$$

Note that the coefficients $\omega_{x,y}(\alpha)$ are unique since the $\vartheta_{\underline{L},y}$ are linearly independent.

Theorem 3.4 *Let $\underline{L} = (L,\beta)$ be a totally positive definite even $\mathcal{O}$-lattice of rank r with level $\mathfrak{l}$.*

(i) The map

$$\omega : \tilde{\Gamma} \to \mathrm{GL}(\mathbb{C}[L^\#/L]), \quad \omega(\alpha)(e_x) := \sum_{y \in L^\#/L} \omega(\alpha)_{y,x} e_y$$

[where $\omega_{y,x}(\alpha)$ *denote the coefficients in* (3.22)*] defines a representation of* $\tilde{\Gamma}$*.*

(ii) The representation ω *is unitary with respect to the scalar product* (2.12)*. It factors through a representation of the finite group* $\tilde{\Gamma}/\Gamma_{\underline{L}}$*, where* $\Gamma_{\underline{L}}$ *is the normal subgroup of* $\tilde{\Gamma}$ *defined in Corollary 3.35.*

(iii) One has

$$\omega(T_b^*)e_x = e\{b\beta(x)\}\, e_x \qquad (b \in \mathcal{O}),$$

$$\omega(S^*)e_x = \sigma(D_{\underline{L}}) \frac{1}{\sqrt{|L^\#/L|}} \sum_{y \in L^\#/L} e\{-\beta(y, x)\}\, e_y,$$

$$\omega(I)e_x = (-1)^r e_x.$$

Proof First of all, we show that for all $\alpha, \alpha' \in \tilde{\Gamma}$, one has

$$\omega(\alpha\alpha') = \omega(\alpha)\omega(\alpha').$$

To prove this identity, it is in fact enough to show

$$\omega(\alpha\alpha')_{y,x} = \sum_{y' \in L^\#/L} \omega(\alpha)_{y,y'} \omega(\alpha')_{y',x}.$$

But since $|_{r/2,\underline{L}}$ defines an action on $\Theta_{\underline{L}}$ (see Theorem 3.1), we easily recognize the above identity. This proves (i).

The fact that ω is unitary follows immediately from Proposition 3.44. This proves the first statement of (ii). The second part of (ii) is immediate by the very definition of $\Gamma_{\underline{L}}$.

Since we have that $\sigma(D_{\underline{L}})$ equals $i^{-nr/2}$ (see Milgram's formula [MH73, p. 127]), part (iii) is immediate by Corollary 3.34. □

Definition 3.49 Let $\underline{L} = (L, \beta)$ be a totally positive definite even $\mathcal{O}$-lattice. Let $\rho : \tilde{\Gamma} \to \mathrm{GL}(V)$ be a finite dimensional representation of $\tilde{\Gamma}$ whose kernel has finite index in $\tilde{\Gamma}$. Let $k \in \frac{1}{2}\mathbb{Z}$. A holomorphic function $F : \mathcal{H} \to V$ satisfying

$$F|_k\alpha = \rho(\alpha)F \quad (\alpha \in \tilde{\Gamma})$$

is called a *vector-valued Hilbert modular form.* Here $\rho(\alpha)F$ denotes that function on $\mathcal{H}$ which at τ in $\mathcal{H}$ takes on the value $\rho(\alpha)\big(F(\tau)\big)$. If $K = \mathbb{Q}$ we require $F(\tau)$ in addition to be bounded on each subset of $\mathcal{H}$ of the form $\Im(\tau) \geq r > 0$. The $\mathbb{C}$-vector space of all such functions is denoted by $M_k(\rho)$.

Let $U := \{b \in \mathcal{O} : \rho(T_b^*) = 1\}$ and $\tilde{U}$ be the dual of U with respect to trace. Then, for any $F \in M_k(\rho)$, we have $F(\tau + b) = F|_k T_b^* = F$, and hence we have a Fourier expansion

$$F(\tau) = \sum_{t \in \tilde{U}} c_F(t)\, q^t \tag{3.23}$$

for suitable $c_F(t) \in V$. Note that, for $K = \mathbb{Q}$, we have $c_F(t) = 0$ unless $t \geq 0$, as follows from the boundedness condition.

Lemma 3.50 (Köcher Principle for Vector-Valued Hilbert Modular Forms) *Suppose $K \neq \mathbb{Q}$ and $F \in M_k(\rho)$. The coefficients $c_F(t)$ in* (3.23) *are equal to zero unless $t \gg 0$ or $t = 0$.*

Proof If e_j $(1 \leq j \leq d)$ is a basis for the space V, we can write

$$F(\tau) = \sum_{j=1}^{d} F_j(\tau)\, e_j.$$

Here the F_j are holomorphic functions for all j. If α lies in the kernel of ρ, then $F = F|_k\alpha = \sum_j F_j|_k\alpha\, e_j$, i.e. for all j, we have $F_j = F_j|_k\alpha$. In other words, F_j is a Hilbert modular form of weight k on the kernel of ρ. By [Fre90, Prop. 4.9] the F_j satisfy the Köcher principle (loc. cit. even weight automorphic forms are considered, but it is easy to modify the proof in loc. cit so that it also covers our case). Therefore, we have

$$F_j(\tau) = \sum_{\substack{t \in \tilde{U} \\ t \gg 0 \text{ or } t = 0}} c_{F_j}(t)\, q^t,$$

and hence

$$F(\tau) = \sum_{\substack{t \in \tilde{U} \\ t \gg 0 \text{ or } t = 0}} \left(\sum_j c_{F_j}(t) e_j \right) q^t.$$

Since $c_F(t) = \sum_j c_{F_j}(t) e_j$, the lemma follows. □

Before we prove the main result of this section, we need a lemma.

Lemma 3.51 *Let V be a finite dimensional G-module, and let ρ be the representation afforded by this G-module. Let v_i $(i = 1, \ldots, n)$ denote a basis for V and define a map $\rho^* : G \to \mathrm{GL}(V)$ by*

$$\rho^*(\alpha) v_i = \sum_{j=1}^{n} \overline{\rho(\alpha)_{ji}}\, v_j.$$

Then ρ^* is a representation of G.

Proof The lemma follows by a straightforward calculation. □

Theorem 3.5 *Let* $\underline{L} = (L, \beta)$ *be a totally positive definite even* $\mathcal{O}$*-lattice of rank* r*, and let* ω *be the representation* (3.22)*. The application*

$$\phi = \sum_{x \in L^\#/L} h_x \vartheta_{\underline{L},x} \mapsto \text{“}\tau \mapsto F(\tau) := \sum_{x \in L^\#/L} h_x(\tau) e_x''$$

defines an isomorphism $\nu : J_{k,\underline{L}}(\chi) \rightarrow M_{k-\frac{r}{2}}(\chi\omega^*)$*. Here* ω^* *denotes the representation associated to* ω *with respect to the basis* e_x $(x \in L^\#/L)$ *as in Lemma 3.51.*

Proof Let $\phi \in J_{k,\underline{L}}(\chi)$. We need to show, first of all, that $\nu(\phi) \in M_{k-\frac{r}{2}}(\chi\omega^*)$. If α is in $\tilde{\Gamma}$, then we have

$$\begin{aligned} \chi(\alpha)\phi = \phi|_{k,\underline{L}}\alpha &= \sum_{x \in L^\#/L} h_x|_{k-r/2}\alpha \; \vartheta_{\underline{L},x}|_{r/2,\underline{L}}\alpha \\ &= \sum_{y \in L^\#/L} \vartheta_{\underline{L},y} \sum_{x \in L^\#/L} h_x|_{k-r/2}\alpha \; \omega(\alpha)_{x,y}. \end{aligned}$$

Since, for fixed τ, the functions $z \mapsto \vartheta_{\underline{L},x}(\tau, z)$ are linearly independent (see Proposition 3.33), we have

$$h_y = \chi(\alpha^{-1}) \sum_{x \in L^\#/L} h_x|_{k-r/2}\alpha \; \omega(\alpha)_{x,y}$$

for $y \in L^\#/L$. Applying α^{-1} to both sides and then writing α for α^{-1} in the resulting identities, we obtain

$$h_y|\alpha = \chi(\alpha) \sum_{x \in L^\#/L} h_x|_{k-r/2} \; \omega(\alpha^{-1})_{x,y}.$$

Using these identities we find

$$F|_{k-r/2}\alpha = \sum_{x \in L^\#/L} h_x|_{k-r/2}\alpha \, e_x = \chi(\alpha) \sum_{y \in L^\#/L} h_y \sum_{x \in L^\#/L} \omega(\alpha^{-1})_{y,x} e_x.$$

Since ω is unitary (see Theorem 3.4) and the e_x form an orthonormal basis, we have $\omega(\alpha^{-1})_{y,x} = \overline{\omega(\alpha)_{x,y}}$ for all x and y in $L^\#/L$. Hence, we have

$$F|_{k-\frac{1}{2}}\alpha = \chi(\alpha) \sum_{y \in L^\#/L} h_y \, \omega^*(\alpha) \, e_y = \big(\chi\omega^*\big)(\alpha)F,$$

which was to be proven.

The injectivity of ν follows from the fact that e_x ($x \in L^\#/L$) form a basis for the space $\mathbb{C}[L^\#/L]$.

Next we prove the surjectivity of ν. Suppose $F \in M_{k-\frac{r}{2}}(\chi\omega^*)$. We need to find some $\phi \in J_{k,\underline{L}}(\chi)$ such that $F = \nu(\phi)$. For each $\tau \in \mathscr{H}$, we have $F(\tau) \in \mathbb{C}[L^\#/L]$. So, we can write $F(\tau) = \sum_{x\in L^\#/L} c_x(\tau)e_x$. Since F is holomorphic, the functions c_x are holomorphic functions on $\mathscr{H}$ for all $x \in L^\#/L$. We set $\phi := \sum_{x\in L^\#/L} c_x\vartheta_{\underline{L},x}$. We obviously have $F = \nu(\phi)$. It remains to show that ϕ is an element of the space $J_{k,\underline{L}}(\chi)$. The invariance under $H(L)$ is obvious from Proposition 3.36, since $H(L)$ acts trivially on $\Theta_{\underline{L}}$. The invariance under $\tilde{\Gamma}$ follows on reversing the arguments of the first part of the proof. We leave the details to the reader. □

For a subgroup Δ of finite index in $\tilde{\Gamma}$, we use $M_k(\Delta, \chi)$ for the space of Hilbert modular forms of weight $k \in \frac{1}{2}\mathbb{Z}$ and character χ on Δ. If χ is trivial, we shortly write $M_k(\Delta)$ for $M_k(\Delta, 1)$.

Corollary 3.52 *We continue with the notations of Theorem 3.5. For any $x \in L^\#/L$, the function h_x lies in $M_{k-\frac{r}{2}}\big(\mathrm{Ker}(\chi\omega^*)\big)$.*

Proof From Theorem 3.5, we have $F(\tau) = \sum_{x\in L^\#/L} h_x(\tau)\, e_x \in M_{k-\frac{r}{2}}(\chi\omega^*)$. Hence, for any $\alpha \in \tilde{\Gamma}$, the following holds true

$$\big(\chi\omega^*\big)(\alpha)F = F|_{k-r/2}\alpha = \sum_{x\in L^\#/L} h_x|_{k-r/2}\, e_x.$$

But this obviously implies that for any $\alpha \in \mathrm{Ker}(\chi\omega^*)$, we have $h_x|_{k-\frac{r}{2}}\alpha = h_x$ which proves the corollary. □

Corollary 3.53 *The space of Jacobi forms is finite dimensional.*

Proof By Theorem 3.5, we have $J_{k,\underline{L}}(\chi) \simeq M_{k-\frac{r}{2}}(\chi\omega^*)$. By Corollary 3.52 the application

$$F = \sum_{x\in L^\#/L} h_x e_x \mapsto (h_x)_{x\in L^\#/L}$$

defines an embedding $M_{k-\frac{r}{2}}(\chi\omega^*) \to \bigoplus_{x\in L^\#/L} M_{k-\frac{r}{2}}\big(\mathrm{Ker}(\chi\omega^*)\big)$. The corollary is now immediate from the subsequent Lemma 3.54. □

Lemma 3.54 *For a subgroup Δ of $\tilde{\Gamma}$ of finite index in $\tilde{\Gamma}$ the dimension of the space of Hilbert modular forms $M_k(\Delta)$ is finite.*

Proof By [Fre90, Thm. 6.1] the space of Hilbert modular forms of even weight is finite. If k is not even, then let ϑ be a Hilbert modular form on some congruence subgroup, say Δ_1, of $\tilde{\Gamma}$ of weight $1/2$, and consider the embedding

$$M_k(\Delta) \to M_{k+3/2}(\Delta \cap \Delta_1), \qquad f \mapsto f\vartheta^3 \qquad \text{if } k \in 1/2 + 2\mathbb{Z},$$
$$M_k(\Delta) \to M_{k+1/2}(\Delta \cap \Delta_1), \qquad f \mapsto f\vartheta \qquad \text{if } k \in 3/2 + 2\mathbb{Z},$$
$$M_k(\Delta) \to M_{k+1}(\Delta \cap \Delta_1), \qquad f \mapsto f\vartheta^2 \qquad \text{if } k \text{ is odd},$$

which in each case implies again that $M_k(\Delta)$ is finite dimensional.

As function ϑ one can take (see Example 3.48) $\vartheta_{\underline{L},0}(\tau, 0)$ for any even $\underline{L}$ of rank one and, which defines a Hilbert modular form on $\Gamma_{\underline{L}}$ (see Corollary 3.35 for $\Gamma_{\underline{L}}$). □

Proof of Theorem 3.2 We write $\phi = \sum_{x \in L^\#/L} h_x \vartheta_{\underline{L},x}$, where, for each x, we have $h_x = \sum_{d \in \beta(x)-h+\mathfrak{d}^{-1}} C(d,x)q^{-d}$, where h is an element of K such that $\chi(T_b) = e\{hb\}$ for $b \in \mathcal{O}$, and where $C(d,x) = c(\beta(x)-d,x)$ (see Theorem 3.3). From Corollary 3.52, the functions h_x are vector-valued Hilbert modular forms. Hence, by Lemma 3.50 we have that $C(d,x) = 0$ unless $d \ll 0$ or $d = 0$, i.e. that $c(t,x) = 0$ unless $t - \beta(x) \gg 0$ or $t = \beta(x)$. This proves the claimed statement. □

Appendix: Jacobi Forms of Odd Index

In this appendix we discuss briefly the notion of Jacobi forms whose index is a not necessarily even $\mathcal{O}$-lattice. Moreover, we shall prove a proposition which links Jacobi forms over number fields with Hilbert modular forms, and which justifies the informal description of Jacobi forms which is given in the introduction.

Let $\underline{L} = (L, \beta)$ denote a (totally positive definite) $\mathcal{O}$-lattice. Assume that $\underline{L}$ is odd, i.e. that β takes on values in $\mathfrak{d}^{-1}$ as for all lattices considered in this treatise, but that there exist elements x in L such that $\beta(x) = \frac{1}{2}\beta(x,x)$ is not in $\mathfrak{d}^{-1}$. Note that for such x there exist a in $\mathcal{O}$ such that $e\{a\beta(x)\} = -1$.

Let k be a half integer and let χ be a linear character of $\tilde{\Gamma}$. If $\underline{L} = (L, \beta)$ is odd, then there is no nonzero function ϕ which satisfies

$$\phi|_{k,\underline{L}}\alpha = \chi(\alpha)\phi \qquad (\alpha \in \tilde{\Gamma}) \tag{3.24}$$
$$\phi|_{k,\underline{L}}h = \phi \qquad (h \in H(L)). \tag{3.25}$$

Indeed, in this case $H(L)$ is not normalized by $\tilde{\Gamma}$. Namely, for $h \in L$, we have $\alpha^{-1}h\alpha \in H(L)\big(0, f_A(x,y)\big)$, where $f_A(x,y) = e\{ab\beta(x) + cd\beta(y)\}$ and (x,y) is the first component of h and $A = \left(\begin{smallmatrix} a & b \\ c & d \end{smallmatrix}\right)$ is the first component of α. Hence, if ϕ satisfies (3.24) and (3.25), then applying α^{-1}, h and α successively to ϕ yields $\phi|(\alpha^{-1}h\alpha) = \phi$ (where we used $|$ for $|_{k,\underline{L}}$). On the other hand, if we write $\alpha^{-1}h\alpha = h'\big(0, f_A(x,y)\big)$, then h' is in $H(L)$ and hence $\phi|(\alpha^{-1}h\alpha) = f_A(x,y)\phi$, so that, since ϕ is different from zero, we have $f_A(x,y) = 1$. But since $\underline{L}$ is odd we can find A and x and y such that $f_A(x,y) = -1$, a contradiction.

Instead we can ask for functions satisfying (3.24) and $\phi|h = \gamma(x,y)\phi(h \in H(L))$ for a linear character γ of $H(L) \simeq L \times L$. If such a character and nonzero ϕ exist then, by a similar reasoning as before, we conclude that

$$\gamma(x,y) = \gamma\big((x,y)A\big)\, e\left\{ab\beta(x) + cd\beta(y)\right\} \qquad (x, y \in L,\ A \in \Gamma). \qquad (3.26)$$

It is not hard to show that the character γ is uniquely determined by these identities, namely, one finds $\gamma(x,y) = e\{\beta(x) + \beta(y)\}$ (see [BS14b]). This function defines a character of $H(L)$, but it does not necessarily satisfy (3.26). It is not obvious when $\gamma(x,y) = e\{\beta(x) + \beta(y)\}$ satisfies (3.26); it does it for instance, if $a^2 + ab + b^2 \equiv 1 \bmod 2$ for all relatively prime elements a and b in $\mathcal{O}$ (which depends on the splitting and ramification of 2 in K). We call a lattice $\underline{L}$ *weakly-odd* if $\gamma(x,y) = e\{\beta(x) + \beta(y)\}$ satisfies the identity (3.26). An even lattice is, of course, weakly-odd.

Definition 3.55 For a totally positive definite, not necessarily even $\mathcal{O}$-lattice we let $J_{k,\underline{L}}(\chi)$ denote the space of holomorphic functions ϕ on $\mathcal{H} \times L_{\mathcal{C}}$ which satisfy (3.24) and

$$\phi|_{k,\underline{L}}h = e\left\{\beta(x) + \beta(y)\right\}\phi \quad \text{for all } h = (x, y, e\left\{\beta(x,y)/2\right\}) \in H(L). \qquad (3.27)$$

Remark Note that, for even $\mathcal{O}$-lattices, this definition coincides with the one given in Sect. 3.6. The discussion above shows that $J_{k,\underline{L}}(\chi) = 0$ unless $\underline{L}$ is weakly odd. Examples of Jacobi forms with weakly-odd index can be found in the next chapter.

We conclude this appendix by a proposition in which a Jacobi form $\phi(\tau, z)$ specialized in the z-variable to *division points* of $L\tau + L$ yields Hilbert modular forms. More precisely, we have:

Proposition 3.56 *Let ϕ be in $J_{k,\underline{L}}(\chi)$, where $\underline{L}$ denotes a not necessarily even $\mathcal{O}$-lattice. We set $\psi(\tau,x,y) := \phi(\tau, x\tau + y)\, e\{\tau\beta(x)\}$. Then we have:*

(i) The function $\psi(\tau,x,y)$ is quasi-periodic in the variables x and y in $\mathcal{R}$ with respect to the $\mathcal{O}$-module L. More precisely, for any λ, μ in L, we have $\psi(\tau, x + \lambda, y + \mu) = e\{-\beta(\lambda, y) + \beta(\lambda + \mu)\}\ \psi(\tau, x, y)$.
(ii) For fixed x and y in K, the map $\tau \mapsto \psi(\tau,x,y)$ defines a Hilbert modular form of weight k and character $\chi\delta_{x,y}$ on the inverse image of $\Gamma(\mathfrak{a}^2)$ in $\tilde{\Gamma}$, where $\mathfrak{a}$ denotes the ideal of all a in $\mathcal{O}$ such that ax and ay are in L, and where $\delta_{x,y}$ is trivial if $\underline{L}$ is even, and trivial or quadratic otherwise.

Remark Note that, for $\underline{L} = (\mathfrak{c}, \omega)$, $\mathfrak{a}$ is the least common multiple of the denominators of $x\mathfrak{c}^{-1}$ and $y\mathfrak{c}^{-1}$.

Proof of Proposition 3.56 We use $\underline{L} = (L, \beta)$ for the $\mathcal{O}$-lattice $(\mathfrak{c}, \omega)$. It is easy to see using Proposition 3.25 that

$$\psi(\tau,x,y) = \big(\phi|_{k,\underline{L}}(x, y, e\left\{-\beta(x,y)/2\right\})\big)(\tau, 0).$$

We prove (i). Using the multiplication law in the Heisenberg group [see (3.4)] and the invariance of ϕ under $H(L)$, we find (writing $|$ for $|_{k,\underline{L}}$)

$$\begin{aligned}
&\psi(\tau, x+\lambda, y+\mu)\\
&= \big(\phi|(x+\lambda, y+\mu, e\{-\beta(x+\lambda, y+\mu)/2\})\big)(\tau, 0)\\
&= e\{-\beta(\lambda, y)\}\,\big(\phi|(\lambda, \mu, \beta(\lambda,\mu)/2)(x, y, e\{-\beta(x,y)/2\})\big)(\tau, 0)\\
&= e\{-\beta(\lambda, y) + \beta(\lambda+\mu)\}\,\big(\phi|(x, y, e\{-\beta(x,y)/2\})\big)(\tau, 0)\\
&= e\{-\beta(\lambda, y) + \beta(\lambda+\mu)\}\,\psi(\tau, x, y).
\end{aligned}$$

Next we prove (ii). Let α be in $\tilde{\Gamma}$, and let A denote the first component of α. Using the multiplication in the Jacobi group and that $\phi|\alpha = \chi(\alpha)\phi$, we have

$$\begin{aligned}
\psi(\tau, x, y)|_k\alpha &= \big(\phi|(x, y, e\{-\beta(x,y)/2\})\alpha\big)(\tau, 0)\\
&= \Big(\phi|\alpha\big((x,y)A, e\{-\beta(x,y)/2\}\big)\Big)(\tau, 0)\\
&= e\big\{[\beta\big((x,y)A\big) - \beta(x,y)]/2\big\}\,\chi(\alpha)\,\psi\big(\tau, (x,y)A\big).
\end{aligned}$$

Now, suppose that x and y are in K. Assume, first of all, that $\underline{L}$ is even. Then by part (i) we see that $\psi(\tau, x, y)$ is periodic with respect to $\mathfrak{a}L \times L$. Assume that A is in $\Gamma(\mathfrak{a}^2)$. We then have $(x,y)A \equiv (x,y) \bmod \mathfrak{a}L \times L$. Moreover, $\beta\big((x,y)A\big)/2 - \beta(x,y)/2 = ab\beta(x) + bc\beta(x,y) + cd\beta(y)$. But each of the three terms on the right is in $\mathfrak{d}^{-1}$. Indeed, to prove, e.g., that $b\beta(x)$ is in $\mathfrak{d}^{-1}$, we write $b = \sum b_j a_j^2$ with numbers b_j in $\mathcal{O}$ and a_j in $\mathfrak{a}$ (Lemma 1.14), so that $b\beta(x) = \sum b_j\beta(a_j x)$, which is in $\mathfrak{d}^{-1}$ since $a_j x$ is in L and $\underline{L}$ is even. It follows that $\psi(\tau, x, y)|\alpha = \chi(\alpha)\,\psi(\tau, x, y)$ as claimed.

If $\underline{L}$ is odd, then ϕ^2 is a Jacobi form of even index $\underline{L}(2) = (L, 2\beta)$. By what we have already proved we conclude that $\psi(\tau, x, y)^2$ is then in $M_{2k}\big(\tilde{\Gamma}(\mathfrak{a}^2), \chi^2\big)$, where $\tilde{\Gamma}(\mathfrak{a}^2)$ denotes the inverse image of $\Gamma(\mathfrak{a}^2)$ in $\tilde{\Gamma}$. It follows that $\psi(\tau, x, y)$ is in $M_k\big(\tilde{\Gamma}(\mathfrak{a}^2), \chi'\big)$, where χ' is a character of $\Gamma(\mathfrak{a}^2)$ whose square equals χ^2. □

Chapter 4
Singular Jacobi Forms

As in the previous chapter, K will denote a totally real number field. Similarly, $\mathcal{O}$, $\mathfrak{d}$ will denote the ring of integers and different of K, respectively. Moreover, we shall use $\Gamma = \mathrm{SL}(2, \mathcal{O})$ and $\tilde{\Gamma}$ for the metaplectic cover of Γ.

In the present chapter we shall study singular Jacobi forms over number fields. The main result of this chapter will be the explicit description of all *singular Jacobi forms* whose indices are totally positive definite rank one $\mathcal{O}$-lattices (see Theorems 4.2 and 4.3). In Sect. 4.1, we shall observe that singular Jacobi forms are in one to correspondence with the one-dimensional $\tilde{\Gamma}$-submodules of the spaces of Jacobi theta functions. In Sect. 4.2, we shall present that the spaces of Jacobi theta functions are isomorphic to the Weil representations associated to certain discriminant modules. Using the results of Sects. 2.4 and 2.5, we shall finally be able to describe explicitly all singular Jacobi forms whose indices are totally positive definite rank one $\mathcal{O}$-lattices. This will be carried out in Sect. 4.4.

4.1 Characterization of Singular Jacobi Forms

In this section, we shall characterize the singular Jacobi forms as the one dimensional $\tilde{\Gamma}$-submodules of Weil representations.

Let $\underline{L} = (L, \beta)$ be a totally positive definite even $\mathcal{O}$-lattice of rank r and ϕ in $J_{k,\underline{L}}(\chi)$. (We refer Definition 3.45 for the space $J_{k,\underline{L}}(\chi)$.) Recall from Definition 3.47 that ϕ is a singular Jacobi form if and only if $c(t, s) = 0$ unless $t = \beta(s)$. Here the $c(t, s)$ are the Fourier coefficients of ϕ as given in Theorem 3.2.

For a linear character χ of $\tilde{\Gamma}$, we define

$$\Theta_{\underline{L}}^{\tilde{\Gamma},\chi} := \{\vartheta \in \Theta_{\underline{L}} : \vartheta|_{r/2,\underline{L}}\alpha = \chi(\alpha)\vartheta \text{ for all } \alpha \in \tilde{\Gamma}\}.$$

H. Boylan, *Jacobi Forms, Finite Quadratic Modules and Weil Representations over Number Fields*, Lecture Notes in Mathematics 2130,
DOI 10.1007/978-3-319-12916-7_4

Clearly, the space $\Theta_{\underline{L}}^{\tilde{\Gamma},\chi}$ is a $\tilde{\Gamma}$-submodule of $\Theta_{\underline{L}}$.

Proposition 4.1 *Let $\underline{L} = (L, \beta)$ be a totally positive definite even $\mathcal{O}$-lattice of rank r and $\phi \in J_{k,\underline{L}}(\chi)$. The following statements are equivalent:*

(i) ϕ *is a singular Jacobi form*
(ii) $k = r/2$
(iii) $\phi \in \Theta_{\underline{L}}$
(iv) $\phi \in \Theta_{\underline{L}}^{\tilde{\Gamma},\chi}$

Proof Recall from Theorem 3.3 that we have the following expansion

$$\phi(\tau, z) = \sum_{x \in L^{\#}/L} h_x(\tau)\vartheta_{\underline{L},x}(\tau, z) \tag{4.1}$$

with

$$h_x(\tau) = \sum_{d \in \beta(x) - h + \mathfrak{d}^{-1}} C(d, x)q^d, \tag{4.2}$$

where $C(d, x) = c(\beta(x) - d, x)$ and the $c(t, x)$ are the Fourier coefficients of ϕ, and where $h \in K$ is such that $\chi(T_b) = e\{hb\}$ $(b \in \mathcal{O})$.

$(i) \implies (ii), (iii)$. Suppose ϕ is a singular Jacobi form. Fix $x \in L^{\#}/L$. Since ϕ is a singular Jacobi form, from (4.2) we have that $h_x(\tau) = C(0, x)$, i.e. h_x is a constant. Hence ϕ is in $\Theta_{\underline{L}}$ and has, in particular, weight $r/2$.

$(ii) \implies (i)$. Suppose $k = r/2$. Fix $x \in L^{\#}/L$. From Corollary 3.52, we have that h_x is a Hilbert modular form of weight $k - r/2 = 0$. From [Fre90, Prop. 4.7], we have that h_x is a constant. From Example 3.48, we have that $\vartheta_{\underline{L},x}$ is a singular Jacobi form (for some subgroup of $\tilde{\Gamma}$). Hence ϕ, being a linear combination of singular Jacobi forms (see (4.1)), is also a singular Jacobi form.

$(iii) \implies (ii)$. Suppose $\phi \in \Theta_{\underline{L}}$. Hence ϕ is a linear combination of forms of weight $r/2$, hence has weight $r/2$.

$(iii) \implies (iv)$. Suppose $\phi \in \Theta_{\underline{L}}$. Since ϕ is in $J_{k,\underline{L}}(\chi)$ we have $\phi|_{k,\underline{L}}\alpha = \chi(\alpha)\phi$ $(\alpha \in \tilde{\Gamma})$. Hence, $\phi \in \Theta_{\underline{L}}^{\tilde{\Gamma},\chi}$.

$(iv) \implies (iii)$. This is obvious. □

4.2 Theta Functions and Weil Representations

The main purpose of this section will be to set up natural isomorphisms between Weil representations associated to certain discriminant modules and the $\tilde{\Gamma}$-modules $\Theta_{\underline{L}}$ of Jacobi theta functions. Moreover, as preparation for the complete decomposition of the $\Theta_{\underline{L}}$ in the next section, we shall translate via these isomorphisms the essential ingredients of the representation theory of the Weil representations to the $\tilde{\Gamma}$-modules $\Theta_{\underline{L}}$.

Let $\underline{L} = (L, \beta)$ be a totally positive definite even $\mathcal{O}$-lattice of rank r. From Theorem 3.1, we know that $\Theta_{\underline{L}}$ is a right $\tilde{\Gamma}$-module. So, the space $\Theta_{\underline{L}}$ equipped with the following $\tilde{\Gamma}$-action

$$(\alpha, \vartheta) \mapsto \vartheta|_{r/2,\underline{L}}\alpha^{-1} \quad (\alpha \in \tilde{\Gamma})$$

becomes a left $\tilde{\Gamma}$-module. This space will be denoted in the following by $\Theta_{\underline{L}}^{\diamond}$. For the definition of the $|_{r/2,\underline{L}}$-action, the reader is referred to Proposition 3.28.

Proposition 4.2 *Let $\underline{L} = (L, \beta)$ be a totally positive definite even $\mathcal{O}$-lattice. The linear continuation of the map*

$$\phi_{\underline{L}} : W(D_{\underline{L}}^{-1}) \to \Theta_{\underline{L}}^{\diamond}, \quad e_{x+L} \mapsto \vartheta_{\underline{L},x}$$

defines a $\tilde{\Gamma}$-linear isomorphism.

Proof Clearly, $\phi_{\underline{L}}$ is a well-defined linear map. From Proposition 3.33 we know that for fixed τ, the functions $z \mapsto \vartheta_{\underline{L},x}(\tau, z)$ $(x \in L^{\#}/L)$ are linearly independent. Hence, $\phi_{\underline{L}}$ is injective. Since $\Theta_{\underline{L}}^{\diamond}$ is spanned by the functions $\vartheta_{\underline{L},x}$ (x in $L^{\#}/L$), the map $\phi_{\underline{L}}$ is also surjective.

It remains to show that $\phi_{\underline{L}}$ is $\tilde{\Gamma}$-linear. Since the group $\tilde{\Gamma}$ is generated by T_b^* $(b \in \mathcal{O})$, S^* and I (see Proposition 3.13), it is enough to prove for those types of elements α, the following identity:

$$\phi_{\underline{L}}(\alpha e_{x+L}) = \vartheta_{\underline{L},x}|_{r/2,\underline{L}}\alpha^{-1}.$$

Applying Theorem 3.1 to the element $(T_b^*)^{-1}$, we see that the claimed identity holds true for $(T_b^*)^{-1}$, since we have

$$\vartheta_{\underline{L},x}|_{r/2,\underline{L}}(T_b^*)^{-1} = e\{-b\beta(x)\}\,\vartheta_{\underline{L},x} = \phi_{\underline{L}}(T_b^* e_{x+L}).$$

Proceeding as in the proof of Corollary 3.34 (ii), we can easily obtain

$$\vartheta_{\underline{L},x}|_{r/2,\underline{L}}(S^*)^{-1} = (-i)^{-nr/2}\frac{1}{\sqrt{|L^{\#}/L|}}\sum_{y \in L^{\#}/L} e\{\beta(y,x)\}\,\vartheta_{\underline{L},y},$$

where r stands for the rank of $\underline{L}$. To prove the claimed identity for $(S^*)^{-1}$, it remains to show that $\sigma(D_{\underline{L}}^{-1}) = (-i)^{-nr/2}$. But this follows from Milgram's formula [MH73, p. 127]. The claimed identity obviously holds true for the element I. □

As preparation for the next section we append here two lemmas.

Lemma 4.3 *Let $\underline{L} = (L, \beta)$ be a totally positive definite even $\mathcal{O}$-lattice, let U be an isotropic submodule of the discriminant module $D_{\underline{L}}$ and let $\underline{L}/U = (\pi^{-1}(U), \beta)$ (see Definition 3.4). Then the following diagram of $\tilde{\Gamma}$-homomorphisms is commutative:*

$$\begin{array}{ccccc}
W\left(D^{-1}_{\underline{L}/U}\right) & \xrightarrow[\underline{\varphi}]{\simeq} & W\left((D^{-1}_{\underline{L}}/U)\right) & \xrightarrow{\phi_{\underline{L}/U}\circ\underline{\varphi}^{-1}} & \Theta^{\diamond}_{\underline{L}/U} \\
 & & \downarrow{\scriptstyle \iota_U} & & \downarrow{\scriptstyle j_U} \\
 & & W\left(D^{-1}_{\underline{L}}\right) & \xrightarrow{\phi_{\underline{L}}} & \Theta^{\diamond}_{\underline{L}}.
\end{array}$$

Here ι_U is the embedding defined in Sect. 2.3, $\phi_{\underline{L}}$ and $\phi_{\underline{L}/U}$ are the isomorphisms from Proposition 4.2, and j_U is the inclusion map. Moreover, $\underline{\varphi}$ denotes the isomorphism induced from the isomorphism φ from Proposition 3.5.

Proof We set $L_1 := \pi^{-1}(U)$. The map $\underline{\varphi}$ is defined by $e_{x+L_1} \mapsto e_{\pi(x)+U}$, where $\pi : L^{\#} \to L^{\#}/L$ is the canonical projection (see Proposition 3.5). To show that the diagram commutes, we need to prove the following identity of maps:

$$j_U \circ \phi_{\underline{L}/U} = \phi_{\underline{L}} \circ \iota_U \circ \underline{\varphi}.$$

On the left we have

$$j_U \circ \phi_{\underline{L}/U}(e_{x+L_1}) = j_U(\vartheta_{\underline{L}/U,x}) = \vartheta_{\underline{L}/U,x} = \sum_{\substack{y \in L_1^{\#}/L \\ y \equiv x \bmod L_1}} \vartheta_{\underline{L},y}.$$

For the last identity we used

$$\vartheta_{\underline{L}/U,x} = \sum_{\substack{r \in L_1^{\#} \\ r \equiv x \bmod L_1}} q^{\beta(r)} e\{\beta(r,z)\}$$

$$= \frac{1}{|L_1|} \sum_{\substack{y \in L_1^{\#}/L \\ y \equiv x \bmod L_1}} \sum_{\substack{r \in L_1^{\#} \\ r \equiv y \bmod L}} q^{\beta(r)} e\{\beta(r,z)\} = \frac{1}{|L_1|} \sum_{\substack{y \in L_1^{\#}/L \\ y \equiv x \bmod L_1}} \vartheta_{\underline{L},y}.$$

On the right we have

$$\phi_{\underline{L}} \circ \iota_U \circ \underline{\varphi}(e_{x+L_1}) = \phi_{\underline{L}} \circ \iota_U(e_{\pi(x)+U}) = \sum_{\substack{Y \in U^{\#}/U \\ Y \equiv \pi(x) \bmod U}} \phi_{\underline{L}}(e_Y)$$

$$= \frac{1}{|L_1|} \sum_{\substack{y \in U^{\#} \\ y \equiv x \bmod L_1}} \vartheta_{\underline{L},y}.$$

For the last identity we did the substitution $Y \mapsto \pi(y)$, where $Y = y + U$. But since $U^{\#} = L_1^{\#}/L$, the diagram commutes.

The map ι_U is $\tilde{\Gamma}$-linear (see Proposition 2.30) and also the maps $\phi_{\underline{L}}$ and $\phi_{\underline{L}/U} \circ \underline{\varphi}^{-1}$ are $\tilde{\Gamma}$-linear (see Proposition 4.2). □

4.3 Decomposition of the $\tilde{\Gamma}$-Modules $\Theta_{\underline{L}}$

In the present section, we shall decompose the spaces of Jacobi theta functions $\Theta_{\underline{L}}$ into irreducible $\tilde{\Gamma}$-submodules, where $\underline{L}$ is a totally positive definite even $\mathcal{O}$-lattice of rank one.

Our main observation is that the discriminant modules $D_{\underline{L}}$ (see Definition 3.3) of such lattices are cyclic finite quadratic $\mathcal{O}$-modules (which follows from Propositions 3.10 and 3.7). The same propositions also imply that if the level of the lattice $\underline{L}$ (the level of $D_{\underline{L}}$) is $\mathfrak{l}$, then the modified level and the annihilator of $\underline{L}$ equals $\mathfrak{l}/4$ and $\mathfrak{l}/2$, respectively.

Definition 4.4 We define the new part $\Theta_{\underline{L}}^{\text{new}}$ of $\Theta_{\underline{L}}$ as the orthogonal complement of $\sum_{U \neq 0} \Theta_{\underline{L}/U}$ with respect to the scalar product (3.18), where U runs through the nonzero isotropic submodules of $D_{\underline{L}}$.

Let $\mathfrak{l}$ and $\mathfrak{a}$ denote the level and annihilator of $\underline{L}$, respectively. Recall that $\mathrm{E}(D_{\underline{L}})$ consists of all $\varepsilon + \mathfrak{a} \in (\mathcal{O}/\mathfrak{a})^*$ such that $\varepsilon \equiv -1 \bmod \mathfrak{h}$ and $\varepsilon \equiv +1 \bmod \mathfrak{a}\mathfrak{h}^{-1}$ for some exact divisor $\mathfrak{h}$ of $\mathfrak{a}$. The group $\mathrm{E}(D_{\underline{L}})$ acts on $\Theta_{\underline{L}}$ via linear continuation of the map $(g, \vartheta_{\underline{L},x}) \mapsto g\vartheta_{\underline{L},x} := \vartheta_{\underline{L},gx}$. Since it acts obviously unitarily and leaves the subspaces $\Theta_{\underline{L}/U}$ invariant (since $\underline{L}/U$ has as the underlying $\mathcal{O}$-module $\pi^{-1}(U)$, where π is the canonical projection from $L^\# \to L^\#/L$, and π is $\mathcal{O}$-linear), it leaves also $\Theta_{\underline{L}}^{\text{new}}$ invariant. For a square-free divisor $\mathfrak{f}$ of $\mathfrak{m}$, we define

$$\Theta_{\underline{L}}^{\text{new},\mathfrak{f}} = \{\vartheta \in \Theta_{\underline{L}}^{\text{new}} : g\vartheta = \psi_{\mathfrak{f}}(g)\vartheta \text{ for all } g \in \mathrm{E}(\underline{M})\}.$$

Here $\psi_{\mathfrak{f}}$ denotes the linear character of $\mathrm{E}(\underline{M})$ such that $\psi_{\mathfrak{f}}(\varepsilon + \mathfrak{a}) = (-1)^t$, where t is the number of primes in $(\mathfrak{h}, \mathfrak{f})$ and $\mathfrak{h}$ is as above (see Proposition 1.23).

Theorem 4.1 *Let $\underline{L} = (L, \beta)$ be a totally positive definite even $\mathcal{O}$-lattice of rank one with annihilator $\mathfrak{a}$, level $\mathfrak{l}$ and modified level $\mathfrak{m}$.*

(i) For every square-free divisor $\mathfrak{f}$ of $\mathfrak{m}$, the space $\Theta_{\underline{L}}^{new,\mathfrak{f}}$ is $\tilde{\Gamma}$-invariant and irreducible.

(ii) One has the following decompositions

$$\Theta_{\underline{L}} = \bigoplus_{\mathfrak{b}^2 \mid \mathfrak{m}} \Theta^{new}_{\mathfrak{a}\mathfrak{b}^{-1}L^\# + L} \tag{4.3}$$

$$\Theta_{\underline{L}}^{new} = \bigoplus_{\substack{\mathfrak{f} \mid \mathfrak{m} \\ \mathfrak{f} \text{ square-free}}} \Theta_{\underline{L}}^{new,\mathfrak{f}}. \tag{4.4}$$

For the proof, which will be a consequence of the decomposition of $W(D_{\underline{L}})$ given in Sect. 2.4, we need a lemma.

Lemma 4.5 *Let $\underline{L}$ be a totally positive definite even $\mathcal{O}$-lattice of rank one whose modified level is $\mathfrak{m}$. Let $\phi_{\underline{L}}$ be the $\tilde{\Gamma}$-module isomorphism in Proposition 4.2. For a square-free divisor $\mathfrak{f}$ of $\mathfrak{m}$, one has*

$$\phi_{\underline{L}}\big(W(D_{\underline{L}}^{-1})^{new}\big) = (\Theta_{\underline{L}}^{\diamond})^{new}, \quad \phi_{\underline{L}}\big(W(D_{\underline{L}}^{-1})^{new,\mathfrak{f}}\big) = (\Theta_{\underline{L}}^{\diamond})^{new,\mathfrak{f}}.$$

Proof Firstly, let $v \in W(D_{\underline{L}}^{-1})^{\text{new}}$. We need to show that $\langle \phi_{\underline{L}}(v), \sum_{U\neq 0} \Theta_{\underline{L}/U}^{\diamond}\rangle = 0$, where U runs through isotropic submodules of $D_{\underline{L}}^{-1}$. Using the fact that the scalar product (2.12) on $W(D_{\underline{L}}^{-1})$ satisfies for all $v, v' \in W(D_{\underline{L}}^{-1})$, $\langle \phi_{\underline{L}}(v), \phi_{\underline{L}}(v')\rangle = \langle v, v'\rangle$ (since $\phi_{\underline{L}}$ is an isomorphism), it is enough then to show $\langle v, \sum_{U\neq 0} \phi_{\underline{L}}^{-1}\Theta_{\underline{L}/U}^{\diamond}\rangle = 0$. But since $\Theta_{\underline{L}/U}^{\diamond} = \phi_{\underline{L}/U} \circ \varphi^{-1}(W(D_{\underline{L}}^{-1}/U))$, and $\phi_{\underline{L}}^{-1} \circ \phi_{\underline{L}/U} \circ \varphi^{-1} = \iota_U$ (see Lemma 4.3), the claimed identity holds true, since v lies in the new part of $W(D_{\underline{L}}^{-1})$.

Secondly, let $v \in (\Theta_{\underline{L}}^{\diamond})^{\text{new}}$. Then we have $\langle v, \sum_{U\neq 0} \Theta_{\underline{L}/U}^{\diamond}\rangle = 0$. But then applying $\phi_{\underline{L}}^{-1}$ (it leaves the scalar product invariant) to this identity and using the two identities in the previous paragraph (which follows from Lemma 4.3), we see that $\phi_{\underline{L}}^{-1}(v)$ must lie in the space $W(D_{\underline{L}}^{-1})^{\text{new}}$, which proves the first identity in the statement of the lemma.

Since $\phi_{\underline{L}}$ is obviously an E($\underline{M}$)-module isomorphism, using the first identity in the statement of the lemma, the second identity holds true. □

Proof of Theorem 4.1 Proof of part (ii). First we prove the identity (4.3). We decompose $W(D_{\underline{L}}^{-1})$ into $\tilde{\Gamma}$-submodules using Theorem 2.5 (i). Applying the isomorphism $\phi_{\underline{L}}$ (which is given in Proposition 4.2) to the decomposition of $W(D_{\underline{L}}^{-1})$ and using Lemma 4.5, we obtain a decomposition of $\Theta_{\underline{L}}^{\diamond}$ into $\tilde{\Gamma}$-submodules as the one in (4.3). But the underlying spaces of $\Theta_{\underline{L}}$ and $\Theta_{\underline{L}}^{\diamond}$ are equal. Hence, the claimed identity holds true.

Next we prove (4.4). We decompose the space $W(D_{\underline{L}}^{-1})^{\text{new}}$ using Theorem 2.5 (ii) into irreducible $\tilde{\Gamma}$-submodules. Again by applying the isomorphism $\phi_{\underline{L}}$ to the decomposition of $W(D_{\underline{L}}^{-1})^{\text{new}}$ and using Lemma 4.5, we obtain a decomposition of $\Theta_{\underline{L}}^{\diamond}$ as of the kind in (4.4). But the underlying spaces of $\Theta_{\underline{L}}$ and $\Theta_{\underline{L}}^{\diamond}$ are the same, hence the claimed identity holds true.

Proof of part (i). Part (i) is an immediate consequence of Lemma 4.5. The fact that the spaces $\Theta_{\underline{L}}^{\text{new},\mathfrak{f}}$ are irreducible follows from the proof of (4.4). □

4.4 The Singular Jacobi Forms of Rank One Index

In this section we shall describe explicitly all singular Jacobi forms whose indices are totally positive definite rank one $\mathcal{O}$-lattices.

Recall that, for even $\underline{L}$, the discriminant modules $D_{\underline{L}}$ (see Definition 3.3) of such lattices are cyclic finite quadratic $\mathcal{O}$-modules (see Propositions 3.10 and 3.7). Recall also from the same propositions that if the level of the lattice $\underline{L}$ (the level of $D_{\underline{L}}$) is $\mathfrak{l}$, then the modified level and the annihilator of $\underline{L}$ equals $\mathfrak{l}/4$ and $\mathfrak{l}/2$, respectively.

From Proposition 3.13, we know that the group $\tilde{\Gamma}$ is generated by the elements T_b^* ($b \in \mathcal{O}$), S^* and I. Hence, the abelianized group $\tilde{\Gamma}^{\mathrm{ab}} = \tilde{\Gamma}/C$ of $\tilde{\Gamma}$ is generated by the elements T_b^*C ($b \in \mathcal{O}$), S^*C and IC. Here C denotes the commutator subgroup of $\tilde{\Gamma}$. But since $(S^*T^*)^3 = (S^*)^2$, and the group $\tilde{\Gamma}^{\mathrm{ab}}$ is abelian, we have that $(S^*)^3C(T^*)^3C = (S^*)^2C$. This implies that $S^*C = (T^*)^{-3}C$, i.e. $\tilde{\Gamma}$ is in fact has a smaller set of generators, namely the elements T_b^*C ($b \in \mathcal{O}$) and IC. Therefore, any character χ of $\tilde{\Gamma}$ is uniquely determined by the value of χ at T_b ($b \in \mathcal{O}$) and at I, since any homomorphism of $\tilde{\Gamma}$ factors through a homomorphism of $\tilde{\Gamma}^{\mathrm{ab}}$.

Recall that an odd character ideal is an integral $\mathcal{O}$-ideal which is a (possibly empty) product of pairwise different prime ideals of degree one over 3.

Definition 4.6 Let $(\mathfrak{c}, \omega)$ be as defined in Definition 3.6. Suppose 2 splits completely in K, and let $\mathfrak{g}$ be an odd character ideal such that $\mathfrak{c}^2\omega\mathfrak{d} = \mathfrak{g}$. We denote by $\varepsilon_{(\mathfrak{c},\omega)}$ that character of $\tilde{\Gamma}$ such that $\varepsilon_{(\mathfrak{c},\omega)}(T_b^*) = e\left\{b\omega\gamma^2/8\right\}$ and $\varepsilon_{(\mathfrak{c},\omega)}(I) = -1$, where $\gamma + 4\mathfrak{c}$ is a generator for $\mathfrak{c}\mathfrak{g}^{-1}/4\mathfrak{c}$.

Remark The fact that $\varepsilon_{(\mathfrak{c},\omega)}$ defines indeed a character of $\tilde{\Gamma}$ is a consequence of Theorem 4.2. Note that the character $\varepsilon_{(\mathfrak{c},\omega)}$ does not factor through a character of Γ, whereas its square does (since it maps $(1, \pm 1)$ to 1). A complete classification of linear characters of Γ including explicit formulas is given in [BS13], which can be used for obtaining explicit formulas for $\pm\varepsilon_{(\mathfrak{c},\omega)}(\alpha)$ for an arbitrary α in $\tilde{\Gamma}$.

If $\mathfrak{g}$ is not one, then the order of $\varepsilon_{(\mathfrak{c},\omega)}$ is 24, and if $\mathfrak{g} = 1$, then it has order 8, as follows from $3\omega\mathfrak{d}\gamma^2 \subseteq 3\omega\mathfrak{d}\mathfrak{c}^2\mathfrak{g}^{-2} = 3\mathfrak{g}^{-1} \subseteq \mathcal{O}$ and the remark preceding Definition 4.6.

We state the three main results of this section.

Theorem 4.2 *Let $\mathfrak{c}$ be a fractional $\mathcal{O}$-ideal and ω a totally positive element in K such that $\mathfrak{c}^2\omega\mathfrak{d} = \mathfrak{g}$ for an odd character ideal $\mathfrak{g}$. Suppose 2 splits completely in K. Set*

$$\vartheta_{(\mathfrak{c},\omega)}(\tau, z) := \sum_{s \in \mathfrak{c}\mathfrak{g}^{-1}} \chi_{4\mathfrak{g}}(s') q^{\frac{1}{8}\omega s^2} e\left\{\omega s z/2\right\}. \tag{4.5}$$

Here $s' \in \mathcal{O}$ is so that $s \equiv s'\gamma \bmod 4\mathfrak{c}$, where $\gamma + 4\mathfrak{c}$ is a generator for $\mathfrak{c}\mathfrak{g}^{-1}/4\mathfrak{c}$ and $\chi_{4\mathfrak{g}}$ is the totally odd Dirichlet character modulo $4\mathfrak{g}$ (see Definition 2.44).

Then $\vartheta_{(\mathfrak{c},\omega)}$ *is a Jacobi form on the full modular group of weight* $1/2$*, index* $(\mathfrak{c},\omega)$ *with character* $\varepsilon_{(\mathfrak{c},\omega)}$*.*

Note that $\vartheta_{(\mathfrak{c},\omega)}$ depends also on the generator γ. However, a different generator changes $\vartheta_{(\mathfrak{c},\omega)}$ only by a sign. Therefore, we suppress the dependency on the choice of γ in the notation.

Theorem 4.3 *Let* $\underline{L} = (L, \beta)$ *be a totally positive definite (not necessarily even)* $\mathcal{O}$*-lattice of rank one. The space* $J_{1/2,\underline{L}}(\chi)$ *is trivial unless* 2 *splits completely in* K*, there is a homomorphism from* $\underline{L}$ *into a lattice* $(\mathfrak{c},\omega)$ *of the kind which occurs in Theorem 4.2, and* $\chi = \varepsilon_{(\mathfrak{c},\omega)}$*. If* 2 *splits completely in* K*, if the map* $\varphi : \underline{L} \to (\mathfrak{c},\omega)$ *is a homomorphism into a lattice* $(\mathfrak{c},\omega)$ *as in Theorem 4.2, and if* $\chi = \varepsilon_{(\mathfrak{c},\omega)}$*, then* $J_{1/2,\underline{L}}(\chi) = \mathbb{C}\cdot\vartheta_{(\mathfrak{c},\omega)}(\tau,\varphi(z))$*. (Here* $\varphi(z)$ *denotes the value at* z *of the* $\mathbb{C}$*-linear extension of* φ *to* $L_{\mathscr{C}}$*.)*

Proposition 4.7 *The number of indices modulo isomorphism which admit a nonzero singular Jacobi form equals* $|\,\mathrm{F}(K)| \cdot |\,\mathrm{Cl}^{+}(K)[2]|$*, where* $\mathrm{F}(K)$ *is the subset of the principal genus containing ideals of the form* $\mathfrak{g}\mathfrak{d}^{-1}$ *with* $\mathfrak{g}$ *an odd character ideal, and where* $\mathrm{Cl}^{+}(K)[2]$ *is the kernel of the squaring map of the narrow class group.*

Proof Let $\mathfrak{J}$ denote the group of fractional $\mathcal{O}$-ideals, and $\mathfrak{P}^{+}$ denote the subgroup of principal $\mathcal{O}$-ideals which have totally positive generators. It is easy to see that the following sequence

$$1 \to \mathrm{Ker}(\varphi) \to \{(\mathfrak{c},\omega) : \omega \gg 0\}/\{(a^{-1},a^{2}) : a \in K^{*}\} \xrightarrow{\varphi} \mathfrak{J}^{2}\mathfrak{P}^{+} \to 1,$$

where $\varphi : (\mathfrak{c},\omega)\{(a^{-1},a^{2}) : a \in K^{*}\} \mapsto \mathfrak{c}^{2}\omega$, is exact. (Here the set of pairs $(\mathfrak{c},\omega)$ of ideals and numbers $\omega \gg 0$ is considered as a group via componentwise multiplication.) Using Theorems 4.2 and 4.3, the number of indices modulo isomorphism which admit a nonzero singular Jacobi form equals $|\,\mathrm{F}(K)| \cdot |\,\mathrm{Ker}(\varphi)|$. We calculate the number of elements in $\mathrm{Ker}(\varphi)$. The following sequence is also exact

$$1 \to \mathrm{Ker}(\phi) \to \mathrm{Ker}(\varphi) \xrightarrow{\phi} \mathrm{Ker}\left(\psi : \mathrm{Cl}(K) \to \mathrm{Cl}^{+}(K)\right) \to 1,$$

where ϕ is the map which maps $(\mathfrak{c},\omega)\{(a^{-1},a^{2}) : a \in K^{*}\}$ to the ideal class of $\mathfrak{c}$. Hence, $|\,\mathrm{Ker}(\varphi)| = |\,\mathrm{Ker}(\phi)| \cdot |\,\mathrm{Ker}(\psi)|$. By direct calculation, we find that the number of elements in $\mathrm{Ker}(\phi)$ equals $[(\mathcal{O}^{*})^{+} : (\mathcal{O}^{*})^{2}]$, where $(\mathcal{O}^{*})^{+}$ denotes the group of totally positive units in K. Therefore, the number that we are looking for is $|\,\mathrm{F}(K)| \cdot [(\mathcal{O}^{*})^{+} : (\mathcal{O}^{*})^{2}] \cdot |\,\mathrm{Ker}(\psi)|$. However, by [vdG88, I. 4], we have that the number $[(\mathcal{O}^{*})^{+} : (\mathcal{O}^{*})^{2}] \cdot |\,\mathrm{Ker}(\psi)|$ equals $|\,\mathrm{Cl}^{+}(K)[2]|$. This proves the proposition. □

The rest of this section is devoted to the proofs of the previously stated theorems.

Proof of Theorem 4.2 The sum in (4.5) can be rewritten in the following way:

$$\begin{aligned}\vartheta_{(\mathfrak{c},\omega)}(\tau,z) &= \sum_{s\in\frac{1}{2}\mathfrak{c}\mathfrak{g}^{-1}/2\mathfrak{c}} \chi_{4\mathfrak{g}}(s') \sum_{\substack{y\in\frac{1}{2}\mathfrak{c}\mathfrak{g}^{-1}\\ y\equiv s \bmod 2\mathfrak{c}}} q^{\frac{1}{2}\omega y^2} e\{\omega yz\}\\ &= \sum_{x\in\frac{1}{2}\mathfrak{c}\mathfrak{g}^{-1}/2\mathfrak{c}} \chi_{4\mathfrak{g}}(s')\, \vartheta_{(2\mathfrak{c},\omega),x}.\end{aligned}$$

But the last identity shows that $\vartheta_{(\mathfrak{c},\omega)}$ is the image of the vector which spans the one-dimensional $\tilde{\Gamma}$-subspace of $W(D_{(2\mathfrak{c},\omega)})$ (see Theorem 2.6, (iii)) under the $\tilde{\Gamma}$-module isomorphism $\phi_{(2\mathfrak{c},\omega)}$ (which is given in Proposition 4.2). Hence, by Proposition 4.1, we have that $\vartheta_{(\mathfrak{c},\omega)}$ is a singular Jacobi form.

Since $\vartheta_{(\mathfrak{c},\omega)}$ is a singular Jacobi form, it transforms under the $\tilde{\Gamma}$-action with a suitable character. Next we show that this character is in fact $\varepsilon_{(\mathfrak{c},\omega)}$. According to the observation about the abelianized group of $\tilde{\Gamma}$ which is explained in the beginning of the present section, it suffices to prove for any $b \in \mathcal{O}$, the following identity

$$\vartheta_{(\mathfrak{c},\omega)}|_{1/2,(\mathfrak{c},\omega)} T_b^* = e\left\{b\omega\gamma^2/8\right\} \vartheta_{(\mathfrak{c},\omega)} \quad (b \in \mathcal{O}), \tag{4.6}$$

since we obviously have $\vartheta_{(\mathfrak{c},\omega)}|_{1/2,(\mathfrak{c},\omega)} I = -\vartheta_{(\mathfrak{c},\omega)}$ (see the action given in Proposition 3.28). On the left of (4.6), we have

$$\vartheta_{(\mathfrak{c},\omega)}(\tau+b,z) = \sum_{s\in\mathfrak{c}\mathfrak{g}^{-1}} \chi_{4\mathfrak{g}}(s') q^{\frac{1}{8}\omega s^2} e\left\{b\omega s^2/8\right\} e\{\omega sz/2\}.$$

If we can show that for every $b \in \mathcal{O}$ and every $s \in \mathfrak{c}\mathfrak{g}^{-1}$ with $(s', 4\mathfrak{g}) = 1$, we have $e\left\{b\omega s^2/8\right\} = e\left\{b\omega\gamma^2/8\right\}$, then (4.6) obviously holds true. Write $s = s'\gamma + 4c$ for $c \in \mathfrak{c}$. Then we have

$$\omega s^2/8 - \omega\gamma^2/8 = \omega(s'\gamma+4c)^2/8 - \omega\gamma^2/8 = \omega s'^2\gamma^2/8 + \omega s'\gamma c + 2\omega c^2 - \omega\gamma^2/8.$$

Note that we have $\omega s'\gamma c \subseteq \omega\mathfrak{c}\mathfrak{g}^{-1}\mathfrak{c} = \mathfrak{d}^{-1}$, and also we have $2\omega c^2 \subseteq 2\omega\mathfrak{c}^2 = 2\mathfrak{g}\mathfrak{d}^{-1}$, since $\mathfrak{c}^2\omega\mathfrak{d} = \mathfrak{g}$ (by the assumption of the theorem). Hence it remains to show that $\omega\mathfrak{d}\gamma^2(s'^2-1)/8$ is integral. Since $\gamma \in \mathfrak{c}\mathfrak{g}^{-1}$, it is enough to show that $\omega\mathfrak{d}(\mathfrak{c}\mathfrak{g}^{-1})^2(s'^2-1)/8$ is integral. But since $\mathfrak{c}^2\omega\mathfrak{d} = \mathfrak{g}$, it suffices to show that $8\mathfrak{g}$ divides $s'^2 - 1$. Let $\mathfrak{q}$ be a prime divisor of 2. Since $\mathfrak{q}$ has degree one, we have $\mathcal{O}/\mathfrak{q}^3 \simeq \mathbb{Z}/8\mathbb{Z}$, and hence the group $(\mathcal{O}/\mathfrak{q}^3)^*$ has exponent 2. Therefore, by the assumption $(s', 4\mathfrak{g}) = 1$, we have $s'^2 \equiv 1 \bmod \mathfrak{q}$. If $\mathfrak{g} = 1$, there is nothing left to prove. Suppose $\mathfrak{g} \neq 1$. Let $\mathfrak{p}$ be a prime divisor of $\mathfrak{g}$. Since $\mathfrak{p}$ has degree one, we have $\mathcal{O}/\mathfrak{p} \simeq \mathbb{Z}/3\mathbb{Z}$, and hence the group $(\mathcal{O}/\mathfrak{p})^*$ has order 2. Again by the same assumption, we have $s'^2 \equiv 1 \bmod \mathfrak{p}$. Therefore, the claimed identity holds true, and thus (4.6) holds true.

To prove that $\vartheta_{(\mathfrak{c},\omega)}$ is of index $(\mathfrak{c},\omega)$, we show that $\vartheta_{(\mathfrak{c},\omega)}$ transforms under the $H(\mathfrak{c})$-action via

$$\vartheta_{(\mathfrak{c},\omega)}|_{1/2,(\mathfrak{c},\omega)}h = e\left\{\omega(x+y)^2/2\right\}\vartheta_{(\mathfrak{c},\omega)}, \tag{4.7}$$

where $h = \left(x, y, e\{\omega xy/2\}\right)$ with x and y in $\mathfrak{c}$. Recall that $H(\mathfrak{c})$ is generated by the elements $\left(x, y, e\left\{\frac{1}{2}\omega xy\right\}\right)$ $(x, y \in \mathfrak{c})$. By applying the action of the Heisenberg group to $\vartheta_{(\mathfrak{c},\omega)}$, we have

$$\vartheta_{(\mathfrak{c},\omega)}|_{1/2,(\mathfrak{c},\omega)}h = e\left\{\tau\omega x^2/2 + \omega xz\right\}\vartheta_{(\mathfrak{c},\omega)}(\tau, z + x\tau + y).$$

By evaluating $\vartheta_{(\mathfrak{c},\omega)}$ at $(\tau, z + x\tau + y)$, we obtain

$$\vartheta_{(\mathfrak{c},\omega)}(\tau, z + x\tau + y) = \sum_{s\in\mathfrak{c}\mathfrak{g}^{-1}} \chi_{4\mathfrak{g}}(s')q^{\frac{1}{8}\omega s^2 + \frac{1}{2}\omega sx} e\{\omega sy/2\}\, e\{\omega sz/2\}.$$

First we show that for every $b \in \mathcal{O}$ and every $s \in \mathfrak{c}\mathfrak{g}^{-1}$ with $(s', 4\mathfrak{g}) = 1$, we have $e\{\omega sy/2\} = e\left\{\omega y^2/2\right\}$. We have

$$\omega sy/2 - \omega y^2/2 = \omega(s'\gamma + 4c)y/2 - \omega y^2/2 = \omega s'\gamma y/2 + 2\omega cy - \omega y^2/2.$$

Since $2\omega cy \subseteq 2\omega\mathfrak{c}^2 = 2\mathfrak{g}\mathfrak{d}^{-1}$ (see the assumption of the theorem, i.e. $\mathfrak{c}^2\omega\mathfrak{d} = \mathfrak{g}$), it is enough to show that the ideal $\left(\omega\mathfrak{d}s'\mathfrak{c}\mathfrak{g}^{-1}\mathfrak{c} - \omega\mathfrak{d}\mathfrak{c}^2\right)/2$ is integral. From the assumption of the theorem (i.e. from $\mathfrak{c}^2\omega\mathfrak{d} = \mathfrak{g}$) again, it suffices to show that $s' - \mathfrak{g}$ is divisible by 2. Let $\mathfrak{q}$ be a prime divisor of 2. By the assumption $(s', 4\mathfrak{g}) = 1$, we have that $\mathfrak{q}$ does not divide s'. Obviously, $\mathfrak{q}$ does not divide $\mathfrak{g}$ either. But since $\mathfrak{q}$ has degree one over 2, we have $\mathcal{O}/\mathfrak{q} \simeq \mathbb{Z}/2\mathbb{Z}$, and hence $\mathfrak{q}|s' - \mathfrak{g}$. Thus, the claimed identity holds true.

Therefore, we have

$$\begin{aligned}\vartheta_{(\mathfrak{c},\omega)}(\tau, z + x\tau + y) &= e\left\{\omega y^2/2\right\}e\left\{-\tau\omega x^2/2\right\}\times\\ &\quad\times\sum_{s\in\mathfrak{c}\mathfrak{g}^{-1}}\chi_{4\mathfrak{g}}(s')q^{\frac{1}{8}\omega(s+2x)^2}e\{\omega sz/2\}\\ &= e\left\{\omega y^2/2\right\}e\left\{-\omega x^2/2\right\}e\{-\omega xz\}\times\\ &\quad\times\sum_{s\in\mathfrak{c}\mathfrak{g}^{-1}}\chi_{4\mathfrak{g}}(s'')q^{\frac{1}{8}\omega s^2}e\{\omega sz/2\},\end{aligned}$$

where $s - 2x \equiv s''\gamma \bmod 4\mathfrak{c}$. But then we have

$$e\left\{\tau\omega x^2/2 + \omega xz\right\}\vartheta_{(\mathfrak{c},\omega)}(\tau, z + x\tau + y) = e\left\{\omega y^2/2\right\}\times \sum_{s\in\mathfrak{c}\mathfrak{g}^{-1}}\chi_{4\mathfrak{g}}(s'')q^{\frac{1}{8}\omega s^2}e\{\omega sz/2\}. \tag{4.8}$$

First note that if x and y in $2\mathfrak{c}$, then we have $s'' \equiv s' \bmod 4\mathfrak{g}$. Indeed, multiplying the congruence $s''\gamma \equiv s'\gamma \bmod 4\mathfrak{c}$ with $\mathfrak{c}^{-1}\mathfrak{g}$ and noting that $\gamma\mathfrak{c}^{-1}\mathfrak{g}$ is integral and relatively prime to $4\mathfrak{g}$ (recall $\mathfrak{c}\mathfrak{g}^{-1} = \mathscr{O}\gamma + 4\mathfrak{c}$), we see that the claimed statement holds true. Then the identity (4.8) becomes

$$\vartheta_{(\mathfrak{c},\omega)}|_{1/2,(\mathfrak{c},\omega)}h = \vartheta_{(\mathfrak{c},\omega)},$$

which proves (4.7).

Next suppose that x and y are not in $2\mathfrak{c}$. If we show that $\chi_{\mathfrak{g}}(s'') = \chi_{\mathfrak{g}}(s')$ and $\chi_4(s'') = \chi_4(s'-2)$, then using also the easily deduced identity $\chi_4(s'-2) = \chi_4(s')e\left\{\omega x^2/2\right\}$, the identity (4.8) becomes

$$\vartheta_{(\mathfrak{c},\omega)}|_{1/2,(\mathfrak{c},\omega)}h = e\left\{(x+y)^2/2\right\}\vartheta_{(\mathfrak{c},\omega)},$$

which proves (4.7), and hence the theorem.

It remains to prove $s'' \equiv s' \bmod \mathfrak{g}$ and $s'' \equiv s'-2 \bmod 4$. Multiplying the congruence $s''\gamma - s'\gamma \equiv -2x \bmod 4\mathfrak{c}$ similarly as above with $\mathfrak{c}^{-1}\mathfrak{g}$, we obtain $s''-s' \equiv -2\mathfrak{g}x\mathfrak{c}^{-1}(\gamma\mathfrak{c}^{-1}\mathfrak{g})^{-1} \bmod 4\mathfrak{g}$. Here note that $x\mathfrak{c}^{-1}$ and $(\gamma\mathfrak{c}^{-1}\mathfrak{g})^{-1}$ are odd. Therefore, the claimed statement holds true. □

For the proof of Theorem 4.3, we need a lemma and a proposition.

Lemma 4.8 *Let $\underline{L} = (L,\beta)$ be a totally positive definite even $\mathscr{O}$-lattice of rank one with level $\mathfrak{l}$. The space $\Theta_{\underline{L}}^{new}$ contains one-dimensional $\tilde{\Gamma}$-submodules if and only if 2 splits completely in K and $\mathfrak{l} = 8\mathfrak{g}$, where $\mathfrak{g}$ is an odd character ideal. If 2 splits completely in K and $\mathfrak{l} = 8\mathfrak{g}$, then $\Theta_{\underline{L}}^{new}$ contains exactly one one-dimensional $\tilde{\Gamma}$-submodule, namely $\Theta_{\underline{L}}^{new,2\mathfrak{g}}$.*

Proof Suppose 2 splits completely in K and $\mathfrak{l} = 8\mathfrak{g}$. By applying Lemma 2.46 to the $\mathscr{O}$-CM $D_{\underline{L}}^{-1}$, we observe that the space $W(D_{\underline{L}}^{-1})^{\text{new},2\mathfrak{g}}$ is the unique one-dimensional $\tilde{\Gamma}$-submodule of $\Theta_{\underline{L}}^{\text{new}}$. Hence, by Lemma 4.5, the space $\Theta_{\underline{L}}^{\text{new},2\mathfrak{g}}$ is the unique one-dimensional $\tilde{\Gamma}$-submodule of $\Theta_{\underline{L}}^{\text{new}}$.

Suppose on the other hand that the space $\Theta_{\underline{L}}^{\text{new}}$ contains one-dimensional $\tilde{\Gamma}$-submodules. From Lemma 4.5 we know that the space $W(D_{\underline{L}}^{-1})^{\text{new}}$ also contains one-dimensional $\tilde{\Gamma}$-submodules. Hence, by Lemma 2.46, $\mathfrak{l}$ must be a character ideal, i.e. $\mathfrak{l} = \mathfrak{g}\mathfrak{h}^3$, where $\mathfrak{g}$ is an odd character ideal and $\mathfrak{h}$ is a (possibly empty) product of pairwise different prime ideals above 2 of degree one and ramification index one. From Propositions 3.10 and 3.7, we know that $\mathfrak{l}$ is divisible by 4, i.e. we have $\mathfrak{h} = 2\mathscr{O}$. But this implies that 2 splits completely in K and $\mathfrak{l} = 8\mathfrak{g}$. □

Proposition 4.9 *Let $\underline{L} = (L,\beta)$ be a totally positive definite even $\mathscr{O}$-lattice of rank one with level $\mathfrak{l}$ and modified level $\mathfrak{m}$. The following statements hold true.*

(i) The space $\Theta_{\underline{L}}$ contains one-dimensional $\tilde{\Gamma}$-submodules if and only if 2 splits completely in K, and $\mathfrak{l} = 8\mathfrak{g}\mathfrak{b}^2$, where $\mathfrak{g}$ is an odd character ideal, and $\mathfrak{b}$ is an integral $\mathscr{O}$-ideal such that $\mathfrak{b}^2|\mathfrak{m}$.

(ii) *The space $\Theta_{\underline{L}}$ contains at most one one-dimensional $\tilde{\Gamma}$-submodule. As a consequence, the space of singular Jacobi forms with index $\underline{L}$ is at most one dimensional.*

Proof Proof of part (i). Suppose that 2 splits completely in K and $\mathfrak{l} = 8\mathfrak{g}\mathfrak{b}^2$. In the following we denote by $\mathfrak{a}$, the annihilator of $\underline{L}$. By Lemma 4.8, the space $\Theta^{\text{new}}_{\mathfrak{a}\mathfrak{b}^{-1}L^{\#}+L}$ contains one-dimensional $\tilde{\Gamma}$-submodules, since the level of $\mathfrak{a}\mathfrak{b}^{-1}L^{\#}+L$ equals $8\mathfrak{g}$. Indeed, the level of $\mathfrak{a}\mathfrak{b}^{-1}L^{\#}+L$ equals the level of the $\mathcal{O}$-FQM $D_{\underline{L}}^{-1}/(\mathfrak{a}\mathfrak{b}^{-1}L^{\#}/L)$ (see Proposition 3.5) which has level $\mathfrak{l}\mathfrak{b}^{-2} = 8\mathfrak{g}$ by Corollary 1.19. Hence, by (4.3), the space $\Theta_{\underline{L}}$ contains one-dimensional $\tilde{\Gamma}$-submodules.

Now suppose that the space $\Theta_{\underline{L}}$ contains a one-dimensional $\tilde{\Gamma}$-submodule, say W. By combining the identities (4.3) and (4.4), we obtain a decomposition of $\Theta_{\underline{L}}$ into irreducible $\tilde{\Gamma}$-submodules. From Proposition 2.16, we have $W \simeq \Theta^{\text{new},\mathfrak{f}}_{\mathfrak{a}\mathfrak{b}^{-1}L^{\#}+L}$ for some square-free divisor $\mathfrak{f}$ of $\mathfrak{m}$, and an integral $\mathcal{O}$-ideal $\mathfrak{b}$ such that $\mathfrak{b}^2$ divides $\mathfrak{m}$. But by Lemma 4.8, we have that 2 must split completely in K, and that the level of $\mathfrak{a}\mathfrak{b}^{-1}L^{\#}+L$ which equals $\mathfrak{l}\mathfrak{b}^{-2}$ (see the above paragraph) must be equal to $8\mathfrak{g}$, which proves (i).

Proof of part (ii). By Theorem 2.6 we have that the space $W(D_{\underline{L}}^{-1})$ contains at most one one-dimensional $\tilde{\Gamma}$-submodule. Hence, by Proposition 4.2, the space $\Theta_{\underline{L}}$ contains at most one one-dimensional $\tilde{\Gamma}$-submodule.

Proposition 4.1 implies that the space of singular Jacobi forms are in one-to-one correspondence with the one-dimensional $\tilde{\Gamma}$-submodules of $\Theta_{\underline{L}}$. Therefore, the space of singular Jacobi forms of index $\underline{L}$ is at most one dimensional. □

Proof of Theorem 4.3 By Proposition 3.10 we have that $\underline{L}$ is isomorphic to an $\mathcal{O}$-lattice of the form $(\mathfrak{c}', \omega')$, where $\mathfrak{c}'^2\omega'\mathfrak{d}$ is integral and $\omega' \gg 0$. Suppose $J_{1/2,\underline{L}}(\chi)$ is non-zero. We set $\underline{L}(2) := (2\mathfrak{c}', \omega')$. Hence, by Proposition 4.1 we have that the space $\Theta_{\underline{L}(2)}$ contains one-dimensional $\tilde{\Gamma}$-submodules (since $J_{1/2,\underline{L}}(\chi)$ can be identified with a subspace of $J_{1/2,\underline{L}(2)}(\chi)$). Proposition 4.9 (i) implies then that 2 splits completely in K and that the level of $\underline{L}(2)$ (which is $8\mathfrak{c}'^2\omega'\mathfrak{d}$) must be equal to $8\mathfrak{g}\mathfrak{b}^2$ for some integral $\mathcal{O}$-ideal $\mathfrak{b}$ whose square divides the modified level of the $\mathcal{O}$-lattice $\underline{L}(2)$. Hence, we have the identity $(\mathfrak{c}'\mathfrak{b}^{-1})^2\omega'\mathfrak{d} = \mathfrak{g}$. However, this implies that $(\mathfrak{c}'\mathfrak{b}^{-1}, \omega')$ is of the kind which occurs in Theorem 4.2. Since $(\mathfrak{c}', \omega')$ obviously embeds into $(\mathfrak{c}'\mathfrak{b}^{-1}, \omega')$, the $\mathcal{O}$-lattice $\underline{L}$ also embeds into $(\mathfrak{c}'\mathfrak{b}^{-1}, \omega')$.

Now we prove that $\chi = \varepsilon_{(\mathfrak{c}'\mathfrak{b}^{-1},\omega')}$. By Theorem 4.2 we have $\vartheta_{(\mathfrak{c}'\mathfrak{b}^{-1},\omega')}$ in $J_{1/2,(\mathfrak{c}'\mathfrak{b}^{-1},\omega')}(\varepsilon_{(\mathfrak{c}'\mathfrak{b}^{-1},\omega')})$. Proposition 4.9 (ii) implies that $\vartheta_{(\mathfrak{c}'\mathfrak{b}^{-1},\omega')}$ spans the space $J_{1/2,(\mathfrak{c}'\mathfrak{b}^{-1},\omega')}(\varepsilon_{(\mathfrak{c}'\mathfrak{b}^{-1},\omega')})$. But this space can be viewed as a subspace of $J_{1/2,\underline{L}}(\chi)$, since $\underline{L}$ can be embedded into $(\mathfrak{c}'\mathfrak{b}^{-1}, \omega')$, i.e. we have the claimed identity.

Assume $\chi = \varepsilon_{(\mathfrak{c},\omega)}$, 2 splits completely in K, and φ denotes a homomorphism from $\underline{L}$ to $(\mathfrak{c}, \omega)$, where $(\mathfrak{c}, \omega)$ is of the kind which occurs in Theorem 4.2. From Theorem 4.2 we know that $\vartheta_{(\mathfrak{c},\omega)}$ is a singular Jacobi form of index $(\mathfrak{c}, \omega)$, and from Proposition 4.9 (ii) we know that the space of Jacobi forms of a given index is at most one-dimensional, hence $J_{1/2,(\mathfrak{c},\omega)}(\chi) = \mathbb{C} \cdot \vartheta_{(\mathfrak{c},\omega)}$. However, from $J_{1/2,(\mathfrak{c},\omega)}(\chi) \subseteq J_{1/2,\varphi(\underline{L})}(\chi)$, we have $J_{1/2,\varphi(\underline{L})}(\chi) = \mathbb{C} \cdot \vartheta_{(\mathfrak{c},\omega)}$ (see also Proposition 4.9 (ii)). Since φ is injective (from Sect. 3.1 we know that every homomorphism between totally positive definite $\mathscr{O}$-lattices is injective), the map $\phi(\tau, z) \mapsto \phi(\tau, \varphi(z))$ defines an isomorphism from $J_{1/2,\underline{L}}(\chi)$ to $J_{1/2,\varphi(\underline{L})}(\chi)$ which finally proves the theorem. □

4.5 Constructing Jacobi Forms of Non-Singular Weight

In this last section we take up the definition of general Jacobi forms of Sect. 3.6 and show how to construct explicit examples of non-singular weight forms. We shall be in particular interested in Jacobi forms of integral weight, even rank one index and trivial character since it is expected that these forms will lift to usual Hilbert modular forms. We shall simply write $J_{k,\underline{L}}^{K}$ for the space of Jacobi forms on $\mathrm{SL}(2, \mathscr{O})$ of weight k, index $\underline{L}$ and trivial character.

We shall show how to generate such Jacobi forms from forms of singular weight using three simple functorial principles. We shall end this section with several concrete examples. However, we know of at least two other methods for constructing examples, whose detailed investigation does not quite fit into the range of this monograph, and, in particular, not into the context of Jacobi forms of singular weight. For the sake of completeness we will also sketch these methods here in the hope that they find interest for doing explicit calculations to obtain further interesting examples of Jacobi forms.

We start with a sketch of the two "non-singular" methods. The first one is the well-known method of averaging some simple function over the Jacobi group $J(\underline{L})$ associated to a lattice $\underline{L} = (L, \beta)$ as in Definition 3.18. As an example we take the constant function 1 and set

$$E_{k,\underline{L}}(\tau, z) := \sum_{g \in J(\underline{L})_1 \backslash J(\underline{L})} 1|_{k,\underline{L}}\, g.$$

Here k is an integer and $\underline{L}$ an even $\mathscr{O}$-lattice, and $J(\underline{L})_1$ is the subgroup of all g in $J(\underline{L})$ such that $1|_{k,\underline{L}}\, g = 1$. Using [Fre90, I. Lemma 5.7] is not hard to show that this infinite series converges for $k \geq 3$ (see also [Boy14]). Moreover, the series is identically zero unless $\mathrm{N}(a)^k = +1$ for all units a in $\mathscr{O}^*$ (since the sum remains unchanged if we replace g by $\left(\begin{smallmatrix} a & \\ & a^{-1} \end{smallmatrix}\right) g$, which multiplies it by $N(a)^k$). One can closely follow the method in [EZ85, § 2, p. 18–20] to calculate the Fourier expansion of $E_{k,\underline{L}}$. The result is

Theorem 4.4 ([Boy14]) *Let $\underline{L} = (L, \beta)$ be an even $\mathscr{O}$-lattice, and let k be an integer. Assume $\mathrm{N}(a)^k = +1$ for all a in $\mathscr{O}^*$. Then, for some constant C, one has*

$$E_{k,\underline{L}} = \vartheta_{\underline{L},0} + C \sum_{\substack{t \in \mathfrak{d}^{-1},\, s \in L^{\#} \\ t - \beta(s) \gg 0}} \mathrm{N}(t - \beta(s))^{k-3/2}\, D_{\underline{L}}(t, s; k-1)\, q^t e\{\beta(s, z)\},$$

where, for any sufficiently large complex σ, we use

$$D_{\underline{L}}(t,s;\sigma) = \sum_{\mathfrak{a}} \frac{\left|\{x \in L/\mathfrak{a}L : \beta(x+s) \equiv \beta(s) - t \bmod \mathfrak{a}\mathfrak{d}^{-1}\}\right|}{\mathrm{N}(\mathfrak{a})^{\sigma}},$$

the sum running over all non-zero integral ideals of K.

The constant C, which depends on k and $\underline{L}$ can be made explicit but is not important here. In the case that $\underline{L}$ is a rank one lattice the Dirichlet series can be further simplified and related to Hecke L-series. For details we refer the reader to [Boy14].

In general, one could also average functions like $q^t \zeta^s$, where t, s are fixed elements in $\mathfrak{d}^{-1}$, $L^{\#}$, respectively, and where we use ζ^s for the function $e\{\beta(s,z)\}$ on $L_{\mathscr{C}}$. This would lead into a theory of Poincaré series, i.e. Jacobi forms which correspond to the functionals $\phi \mapsto c_{\phi}(t,s)$ ((t,s)th coefficient of ϕ) if we identify $J_{k,\underline{L}}^{K}(\chi)$ with its dual via a suitable Petersson type scalar product (see [GKZ87, II.2]) for the case $K = \mathbb{Q}$ and scalar index). However, we shall not pursue this here. Arithmetically interesting are those series $E_{k,\underline{L};t,s}$ which we obtain by averaging $q^t\zeta^s$ with $t = \beta(s)$, i.e. the averages $E_{k,\underline{L};\beta(s),s}$ of $q^{\beta(s)}\zeta^s$ where s runs through the elements in $L^{\#}$ such that $\beta(s) \in \mathfrak{d}^{-1}$. One would call these series Jacobi Eisenstein series since they represent the functionals $\phi \mapsto c_{\phi}(\beta(s),s)$ which vanish identically on the subspace of cusp forms, i.e. they would be perpendicular to the space of cusp forms with respect to a Petersson type scalar product. They are arithmetically interesting since the condition $t = \beta(s)$ entails that the Fourier coefficients are special values of Dirichlet series of the type encountered in the discussion of $E_{k,\underline{L}} = E_{k,\underline{L};0,0}$. This can be seen by mimicking the computation of the Fourier coefficients of the Eisenstein series in [EZ85, §2] (or see [Boy14]). We shall not pursue this any further here since the calculation of the Fourier coefficients seems to be tedious and will eventually become easier with the help of a suitable Hecke theory, which has still to be developed.

It is not difficult to count the number of Eisenstein series $E_{k,\underline{L};\beta(s),s}$. Indeed it is easy to see that the average of $q^{\beta(s)}\zeta^s$ over $J(\underline{L})$ is the same as the average of $q^{\beta(s)}\zeta^s|_{k,\underline{L}}g$, where $g = (1,x,0,1)$ or $g = (\left(\begin{smallmatrix} a & \\ & a^{-1}\end{smallmatrix}\right),0,0,1)$ ($x \in L$, $a \in \mathscr{O}^*$). In other words, one has $E_{k,\underline{L};\beta(s),s} = E_{k,\underline{L};\beta(s+x),s+x}$ and $E_{k,\underline{L};\beta(s),s} = \mathrm{N}(a)^k E_{k,\underline{L};\beta(as),as}$ (see Propositions 3.25, 3.28). Thus the dimension of the space of Eisenstein series of the form $E_{k,\underline{L};\beta(s),s}$ ($s \in L^{\#}$, $\beta(s) \in \mathfrak{d}^{-1}$) is bounded to above by the number of orbits

$$\mathscr{O}^*\backslash\{s + L \in L^{\#}/L : \beta(s) \in \mathfrak{d}^{-1},\ \mathrm{N}(\mathscr{O}_s^*)^k = 1\}, \tag{4.9}$$

where $\mathscr{O}_s^*$ is the stabilizer in $\mathscr{O}^*$ of $s + L$ in $L^{\#}/L$. The condition $\mathrm{N}(\mathscr{O}_s^*)^k = 1$ is, of course, void if k is even or if K does not possess units with norm equal to -1. The dimension of the space spanned by the Eisenstein series is even equal to this number of orbits. Namely, following the proof of [EZ85, Thm. 2.4], one finds that

$$E_{k,\underline{L};\beta(s),s} = \text{const.} \sum_{a \in \mathcal{O}^*/\mathcal{O}_s^*} \mathrm{N}(a)^k \vartheta_{\underline{L},as} + \cdots .$$

Here "$\cdots$" indicates that part of the Fourier development of $E_{k,\underline{L};\beta(s),s}$ consisting only of terms $q^t\zeta^s$ with $t-\beta(s) \gg 0$, and const. is a nonzero number. The "singular parts" of the $E_{k,\underline{L};\beta(s),s}$ (s running through a complete set of representatives R for the orbits in (4.9)) are linearly independent as follows from the linear independence of the $\vartheta_{\underline{L},x}$ (Proposition 3.33). Consequently, the $E_{k,\underline{L};\beta(s),s}$ ($s \in R$) are linearly independent too. If K has class number one and $\underline{L}$ has rank and modified level one, then there is only one Eisenstein series, namely, $E_{k,\underline{L},0,0} = E_{k,\underline{L}}$, i.e. the one discussed in the preceding theorem.

The transformation law in Definition 3.45 of a Jacobi form ϕ in $J_{k,\underline{L}}^K$ applied to a diagonal matrix in $\mathrm{SL}(2,\mathcal{O})$ implies that its Fourier coefficients $c(t,s)$ satisfy $c(a^2t, as) = N(a)^k c(t,s)$ for all units a in $\mathcal{O}$. Furthermore, we know from Theorem 3.3 that $c(t,s)$ depends only on $\beta(s) - t$ and $s + L$. Therefore, if the class number of K is one, we find that ϕ is a cusp form if and only if $c(\beta(s), s) = 0$ for all s in a complete system of representatives for the orbits in (4.9). (Here we also use the remark after Definition 3.47). Summarizing, we have proved

Proposition 4.10 *Let $\underline{L}$ be an even $\mathcal{O}$-lattice, $k \geq 3$ be an integer, and suppose that the class number of K is one. Then the co-dimension of the subspace of cusp forms in $J_{k,\underline{L}}^K$ equals the cardinality of the set of orbits in* (4.9).

If the class number of K is strictly greater than one we have to take into account more than one cusp, which can be done by considering the average of functions $q^t\zeta^s|_{k,\underline{L}}g$, where again t is in $\mathfrak{d}^{-1}$, s is in $L^\#$ such that $t = \beta(s)$, and where now g is an arbitrary element in $\mathrm{SL}(2,K)$. We can proceed as before to count the co-dimension of the subspace of cusp forms in $J_{k,\underline{L}}^K$. This is not difficult but the resulting formulas will become more tedious.

The second method uses the fact that the lth Taylor coefficient

$$\phi_l(\tau) = \frac{\partial^{l_1+\cdots+l_m}}{\partial z_1^{l_1}\cdots\partial z_m^{l_m}}\phi(\tau,z)|_{z=0} \qquad (l = (l_1,\ldots,l_m))$$

of a Jacobi form ϕ of index $\underline{L}$ (with respect to any given system of coordinate functions z_j on $L_{\mathscr{C}}$) in the Taylor expansion of a Jacobi form $\phi(\tau,z)$ around $z=0$ is not an arbitrary holomorphic function of τ in $\mathscr{H}$. Indeed, $\phi_0(\tau)$ equals the nullwert $\phi(\tau,0)$ of ϕ, and the transformation formulas for Jacobi forms imply that this is a Hilbert modular form on the full Hilbert modular group of the same weight as ϕ. For general l one obtains functions which have a transformation law under $\mathrm{SL}(2,\mathcal{O})$ similar to the quasi-modular forms in the theory of elliptic modular forms. These are distinguished functions equal to or generalizing usual Hilbert modular forms, and, for certain K, the spaces spanned by these functions can be explicitly described. Since $\phi(\tau,z)$ as a function of z is multivariate one would not a priori expect that finitely many coefficients ϕ_l suffice to recover ϕ. However, if one picks a finite set S of $r = \text{rank}(\underline{L})$ many multi-indices which satisfies

If l is in S then all $h \in \mathbb{Z}^m_{\geq 0}$ with $h \leq l$ and $h \equiv l \bmod 2$ are contained in S

then, under certain additional conditions, the map $\phi \mapsto \{\phi_l\}_{l \in S}$ is injective. (The relations "$\leq$" and "$\equiv$" have to be read componentwise.) Even more, there is an explicit finite closed formula to express any ϕ of a given $\underline{L}$ in terms of the finitely many Taylor coefficients ϕ_l (l in S). The "certain additional conditions" are that $W(D_{\underline{L}})$ is irreducible and that the Wronski type determinant

$$\det\left((\vartheta_{\underline{L},x_i})_{l^{(j)}}\right)_{1\leq i,j\leq r}$$

does not vanish identically (here $\{x_1, \ldots, x_r\}$ is a complete system of representatives for $L^{\#}/L$ and $S = \{l^{(1)}, \ldots, l^{(r)}\}$, and, as before, $(\vartheta_{\underline{L},x_i})_l$ denotes the lth Taylor coefficient of $\vartheta_{\underline{L},x_i}$ around 0 with respect to the coordinates z_j). Note that $W(D_{\underline{L}})$, for a rank one lattice $\underline{L}$, is irreducible if and only if its modified level is one (see Theorem 2.5). A rank one lattice $(\mathfrak{c}, \omega)$ has modified level $\mathfrak{c}^2\omega\mathfrak{d}/2$ (Prop. 3.7), and thus there exists rank one lattices of modified level one if and only if the different lies in the principal genus of K.

The method of the preceding paragraph can be extended (by passing to certain natural subspaces of $J^K_{k,\underline{L}}(\chi)$), so to dispose of the assumption that $W(D_{\underline{L}})$ is irreducible. For details the reader is referred to [BHS14], where the indicated method is described in detail and explicit working examples are given. For the (classical) case $K = \mathbb{Q}$ and scalar index this method was developed in [Sko85, §7]. The very starting point for this method are the considerations in [EZ85, §3].

We finally explain the three general principles mentioned in the introduction. The first one is multiplication of Jacobi forms. However, since the indices of the Jacobi forms in question might be different, we have to be a bit more precise. We define the product of two Jacobi forms ϕ_1 and ϕ_2 of index $\underline{L}_j = (L_j, \beta_j)$ ($j = 1, 2$), respectively, as

$$(\phi_1 \otimes \phi_2)(\tau, (z_1, z_2)) = \phi_1(\tau, z_1)\phi_2(\tau, z_2) \qquad (z_j \in (L_j)_{\mathscr{C}}). \tag{4.10}$$

The following proposition is immediate from the Definition 3.45 of Jacobi forms.

Proposition 4.11 *The multiplication* (4.10) *defines a map*

$$\otimes : J^K_{k_1,\underline{L}_1}(\chi_1) \times J^K_{k_2,\underline{L}_2}(\chi_2) \to J^K_{k_1+k_2,\underline{L}_1\oplus\underline{L}_2}(\chi_1\chi_2).$$

Here $\underline{L}_1 \oplus \underline{L}_2$ denotes the orthogonal sum *of $\underline{L}_1$ and $\underline{L}_2$ (i.e. the lattice given by $(L_1 \times L_2, (x, y) \mapsto \beta_1(x) + \beta_2(y))$).*

The described multiplication is not always interesting since it increases the rank of the index lattice. One can combine it with the next principle for reducing again the resulting rank. The second principle is a pullback in the lattice-variable. More precisely, assume that $\alpha : \underline{L} \to \underline{L}'$ is an isometric map from the lattice $\underline{L} = (L, \beta)$

into a lattice $\underline{L}' = (L', \beta')$, i.e. a homomorphism of groups $\alpha : L \to L'$ such that $\beta'(\alpha(x)) = \beta(x)$ for all x in L. If ϕ is a Jacobi form of index $\underline{L}'$, we define a function on $\mathscr{H} \times L_{\mathscr{C}}$ by setting

$$(\alpha^*\phi)(\tau, z) := \phi(\tau, \alpha(z)). \tag{4.11}$$

(In this formula α is linearly continued to a map from $L_{\mathscr{C}}$ to $L'_{\mathscr{C}}$). Again the following proposition is an immediate consequence of Definition 3.45.

Proposition 4.12 *For any isometric map $\alpha : \underline{L} \to \underline{L}'$, the application* (4.11) *defines a map*

$$\alpha^* : J^K_{k,\underline{L}'}(\chi) \to J^K_{k,\underline{L}}(\chi). \tag{4.12}$$

We can combine the two described principles as follows. For a positive integer a set $\underline{L}(a) = (L, a\beta)$ and $\oplus^a \underline{L} = \underline{L} \oplus \cdots \oplus \underline{L}$ the a-fold orthogonal sum of $\underline{L}$ with itself. The diagonal map $\Delta : \underline{L}(a) \to \oplus^a \underline{L}$ which maps x in L to $(x, \ldots, x)$ is isometric. The composition of Δ with the product $\otimes$ defines a map from $J^K_{k,\underline{L}}(\chi)$ to $J^K_{k,\underline{L}(2)}(\chi)$. This can, of course, be iterated, so to obtain the map

$$J^K_{k,\underline{L}}(\chi) \times \cdots \times J^K_{k,\underline{L}}(\chi) \to J^K_{k,\underline{L}(a)}(\chi)$$

which, in explicit terms, is nothing else but the simple application which maps Jacobi forms $\phi_1, \ldots, \phi_a$ of the same index $\underline{L}$ to $\phi_1(\tau, z) \cdots \phi_a(\tau, z)$.

The last principle varies the underlying number field by a pullback in the upper half plane variable of a Jacobi form. For this let $K_1 \subseteq K$ be a subfield of K with ring of integers $\mathscr{O}_1$. Note that K_1 is then totally real too (since every embedding of K_1 into $\mathbb{C}$ can be continued to an embedding of K into the complex numbers). We then have $\mathscr{C}_1 := \mathbb{C} \otimes_{\mathbb{Q}} K_1 \subseteq \mathscr{C} = \mathbb{C} \otimes_{\mathbb{Q}} K$, and accordingly $\underline{L}_{\mathscr{C}_1} := \underline{L} \otimes_{\mathscr{O}_1} \mathscr{C}_1 \subseteq \underline{L}_{\mathscr{C}}$. If ϕ is a Jacobi form in $J^K_{k,\underline{L}}(\chi)$, we can restrict ϕ to a function ϕ_1 on $\mathscr{H}_1 \times L_{\mathscr{C}_1}$, where $\mathscr{H}_1$ denotes the set of elements in $\mathscr{C}_1$ with positive imaginary part (see Sect. 3.2). If we examine the definition of the Jacobi group actions 3.24 and 3.25 we find that ϕ_1 transforms like a Jacobi form over K_1 of index

$$\mathrm{Res}_{\mathscr{O}_1} \underline{L} := \left(L, \mathrm{tr}_{K/K_1} \circ \beta\right),$$

respectively. However, in the transformation law 3.24 under $A = \left(\begin{smallmatrix} a & b \\ c & d \end{smallmatrix}\right)$ in $\mathrm{SL}(2, \mathscr{O}_1)$ the factor $\pm \mathrm{N}(c\tau + d)^{-k}$ with τ in $\mathscr{H}_1$ shows up, where N is the norm of the $\mathbb{C}$-algebra $\mathscr{C}$ and not the norm N_1 of $\mathscr{C}_1$. It is quickly checked that, for τ in $\mathscr{H}_1$, one has $\mathrm{N}(\tau) = \mathrm{N}_1(\tau)^d$ with $d = [K : K_1]$ (since the infinite places of K_1 in the totally real extension K do not ramify). We have thus the group homomorphism

$$\iota : \tilde{\Gamma}_1 \to \tilde{\Gamma}, \qquad (A, w) \mapsto (A, w^d)$$

of the metaplectic cover $\tilde{\Gamma}_1$ of $\mathrm{SL}(2,\mathcal{O}_1)$ into the metaplectic cover of $\Gamma = \mathrm{SL}(2,\mathcal{O})$. Finally, for a character χ of $\tilde{\Gamma}$, we use $\mathrm{Res}_{\mathcal{O}_1}(\chi)$ for its composition with ι. The following proposition is now clear.

Proposition 4.13 *Let K_1 be a subfield of K, and let $d = [K : K_1]$. The application which maps a Jacobi form over K of index $\underline{L}$ to its restriction to $\mathcal{H}_1 \times L_{\mathcal{C}_1}$ defines a map*

$$\mathrm{Res}_{K_1} : J^K_{k,\underline{L}}(\chi) \to J^{K_1}_{dk,\mathrm{Res}_{\mathcal{O}_1}\underline{L}}\big(\mathrm{Res}_{\mathcal{O}_1}(\chi)\big).$$

Note that the map Res_{K_1} increases the weight. We can apply the restriction map for example to the singular forms of Theorem 4.2 for obtaining non-singular forms over subfields of K.

Example 4.14 Let $K = \mathbb{Q}(\sqrt{D})$ be the real quadratic field of discriminant D. Assume that $D \equiv 1 \bmod 8$ (i.e. that 2 splits completely in K), and that the fundamental units have norm -1. The first examples of discriminant D fulfilling these assumptions are $D = 17, 41, 65, 73, 89, 97, \ldots$.

Let ε denote the fundamental unit with $\varepsilon > 1$. Here, for an a in K we use a' for its conjugate. The number $\omega := \varepsilon/\sqrt{D}$ is totally positive (since $\varepsilon' = -1/\varepsilon < 0$) and $\omega\mathfrak{d} = 1$. Therefore the pair $(\mathcal{O},\omega)$ defines a totally positive rank one lattice which satisfies the assumption of Theorem 4.2. Hence we have available the singular Jacobi form $\vartheta_{(\mathcal{O},\omega)}$. Note that the unique totally odd Dirichlet character modulo 4 is given by $\big(\frac{-4}{\mathrm{N}(a)}\big)$. We therefore have

$$\vartheta_{(\mathcal{O},\omega)}(\tau,z) = \sum_{s\in\mathcal{O}} \Big(\frac{-4}{\mathrm{N}(s)}\Big) q^{\frac{1}{8}\omega s^2} e\{\omega s z/2\} \in J^{\mathbb{Q}(\sqrt{D})}_{\frac{1}{2},(\mathcal{O},\omega)}\big(\varepsilon_{(\mathcal{O},\omega)}\big).$$

For obtaining forms of non-singular weight with trivial character one can take powers $\vartheta^{\otimes k}_{(\mathcal{O},\omega)}$ with k divisible by 8 (recall from the remark following Definition 4.6 that $\varepsilon_{(\mathcal{O},\omega)}$ has order 8). This gives forms of even weight $k/2$, however, with index $\underline{L} := (\mathcal{O}^k, \omega(x_1y_1 + \cdots + x_ky_k))$, which is odd and of rank k. For obtaining forms with even rank one index let $a_1, \ldots, a_k$ be elements of $\mathcal{O}$ such that $2m := a_1^2 + \cdots + a_k^2$ is divisible by 2. Then $x \mapsto (a_1,\ldots,a_k)x$ defines an isometric map of the lattice $(\mathcal{O}, 2m\omega)$ into $\underline{L}$. Note that the lattice $(\mathcal{O}, 2m\omega)$ is even of modified level (m). Applying Proposition 4.12 we see that

$$\vartheta_{\mathcal{O},\omega}(\tau,a_1z)\cdots\vartheta_{\mathcal{O},\omega}(\tau,a_kz) \in J^{\mathbb{Q}(\sqrt{D})}_{k/2,(\mathcal{O},2m\omega)}.$$

For improving the preceding method one can think of dividing the products considered in this example by suitable Hilbert modular forms to decrease the weight. However, this would need a careful analysis of the zeros of such products which seems to be not obvious. Another possibility is to consider also singular Jacobi forms on the full modular group whose index has rank greater than one and which are not just products of singular forms of rank one like the functions $\vartheta^{\otimes k}_{(\mathcal{O},\omega)}$ of the preceding example. The next example constructs singular Jacobi forms of higher rank index.

Example 4.15 Assume that the narrow ideal class of the different $\mathfrak{d}$ of K is trivial, i.e. that $\mathfrak{d}^{-1} = (\omega)$ for some $\omega \gg 0$. We have then the even rank one lattice $\underline{L} = (\mathscr{O}, 2\omega)$; its dual is $\frac{1}{2}\mathscr{O}$. The $\tilde{\Gamma}$-module $\Theta_{\underline{L}}$ spanned by the theta functions $\vartheta_{\underline{L},x/2}$ $(x \in \mathscr{O}/2\mathscr{O})$ has dimension $d = 2^{[K:\mathbb{Q}]}$. To construct a Jacobi form on the full Hilbert modular group $\Gamma = \mathrm{SL}(2,\mathscr{O})$ we consider the $\tilde{\Gamma}$-module $\bigwedge^d \Theta_{\underline{L}}$. It is one-dimensional, spanned by

$$\phi_K := \sum_\sigma \mathrm{sign}(\sigma)\, \vartheta_{\underline{L},\sigma(x_1)/2} \otimes \cdots \otimes \vartheta_{\underline{L},\sigma(x_d)/2},$$

where x_j $(j = 1, \cdots, d)$ runs through a complete set of representatives for $\mathscr{O}/2\mathscr{O}$, where σ runs through the permutations of $\mathscr{O}/2\mathscr{O}$, and where sign denotes the non-trivial linear character (signature) of this group of permutations. (We allow us here the slight abuse of language to write $\sigma(x_j)$ for the representative $x_{j'}$ of $\sigma(x_j + 2\mathscr{O})$.) From Proposition 4.11 (which is, of course, also valid for Jacobi forms on subgroups of $\tilde{\Gamma}$) we see that ϕ_K is a (singular) Jacobi form of weight $d/2$ on the full modular group. Its character χ maps $\left(\begin{smallmatrix}1 & b\\ 0 & 1\end{smallmatrix}\right)$ to $e\{\omega b N/4\}$, where $N = x_1^2 + \cdots + x_d^2$; it is trivial if $N \equiv 0 \bmod 4$ (cf. Sect. 4.4), which is for example the case if 2 is a prime in K and $d \geq 4$ as in the subsequent example (since then, by an obvious argument, $(a^2-1)N \equiv 0 \bmod 4$ for any unit a modulo 2, and $a^2 - 1$ is a unit modulo 4 if $a \not\equiv 1 \bmod 2$ and $\mathscr{O}/2\mathscr{O}$ is a field). Summing up we have proved

$$\phi_K \in J^K_{d/2, \oplus^d \underline{L}}(\chi) \qquad (d = 2^{[K:\mathbb{Q}]}).$$

We can rewrite the defining equation of ϕ_K in the form

$$\begin{aligned}&\phi_K(\tau, z_1, \ldots, z_d)\\ &\quad = \sum_{r_1,\ldots,r_d \in \mathscr{O}} \mathrm{sign}\left(\begin{smallmatrix}x_1 & x_2 & \cdots & x_d\\ r_1 & r_2 & \cdots & r_d\end{smallmatrix}\right) q^{\omega(r_1^2+\cdots+r_d^2)/4}\, e\left\{\omega(r_1 z_1 + \cdots + r_d z_d)\right\}.\end{aligned}$$

Here the matrix argument of sign stands for the application of $\mathscr{O}/2\mathscr{O}$ to itself which maps $x_1 + 2\mathscr{O}$ to $r_1 + 2\mathscr{O}$, $x_2 + 2\mathscr{O}$ to $r_2 + 2\mathscr{O}$ etc., and where we set $\mathrm{sign}(P) = 0$ if the matrix P does not define a permutation. One can now apply Propositions 4.12, 4.13 to pull back and restrict ϕ_K and obtain forms of non-singular weight and index of rank one.

Example 4.16 We continue the preceding example by the first totally real number field different from $\mathbb{Q}$, which is $K = \mathbb{Q}(\sqrt{5})$ (if one orders number fields by degree and discriminant). Here we can choose $\omega = \varepsilon/\sqrt{5}$, where $\varepsilon = \frac{1+\sqrt{5}}{2}$. As representatives for $\mathscr{O}/2\mathscr{O}$ we can choose $0, 1, \varepsilon, \varepsilon'$. For any a, b, c, d in $\mathscr{O} = \mathbb{Z}[\varepsilon]$ we obtain, using the pullback method, the Jacobi form

$$\begin{aligned}
&\phi_{\mathbb{Q}(\sqrt{5})}(\tau, az, bz, cz, dz) \\
&\quad = \sum_{r,s,t,u \in \mathbb{Z}[\varepsilon]} \operatorname{sign}\left(\begin{smallmatrix} 0 & 1 & \varepsilon & \varepsilon' \\ r & s & t & u \end{smallmatrix}\right) q^{\omega(r^2+s^2+t^2+u^2)/4}\, e\{\omega(ra+sb+tc+ud)z\} \\
&\qquad\qquad\qquad\qquad \in J^{\mathbb{Q}(\sqrt{5})}_{2,(\mathcal{O},2\omega m)} \qquad (m = a^2+b^2+c^2+d^2).
\end{aligned}$$

Depending on the choice of a, b, c, d this form might vanish identically. We leave it to the reader to verify that, for e.g. $a, b, c, d = 0, 1, \varepsilon, \varepsilon'$, we obtain a non-zero form, whose singular part equals $\vartheta_{\underline{L},1} + \vartheta_{\underline{L},\varepsilon} + \vartheta_{\underline{L},\varepsilon'}$ ($\underline{L} = (\mathcal{O}, 8\omega)$), and whose index has modified level 4.

It is clear that these examples, though at the first glance somewhat more "ad hoc" than conceptual, are not at all exceptional. They deserve eventually their proper theory. In any case, one can use these methods to produce a plethora of Jacobi forms over number fields from the singular forms considered in this monograph or from their generalizations to higher rank.

Appendix

Tables

This chapter contains tables which list the first number fields (ordered by increasing discriminant) of degrees 2, 3 and 4 which admit nonzero singular Jacobi forms. More precisely, we searched all of the Bordeaux number field tables of the PARI group [Bor95] for totally real number fields fulfilling the conditions for admitting nonzero singular Jacobi forms.

For number fields K of degree 2 and 3 over $\mathbb{Q}$ we list the first 30 number fields where we find nonzero singular Jacobi forms (Tables A.1 and A.2). The percentage of number fields of degree 2 and 3 admitting nonzero singular Jacobi forms among all fields of these degrees in the Bordeaux tables is 17.87 % and 4.75 %, respectively. For number fields K of degree 4 over $\mathbb{Q}$ we list *all* number fields of the Bordeaux tables admitting nonzero singular Jacobi forms (Table A.3). The corresponding percentage is in this case 0, 25 %. We searched the Bordeaux tables also for number fields of degrees $n = 5, 6, 7$ which admit nonzero singular forms. However, in the available range we could not find any such fields.

The columns of the tables display from left to right the discriminant D_K of K, the number s of nonzero singular Jacobi forms modulo isomorphism, the minimal polynomial f of K, whether the different $\mathfrak{d}_K$ of K is a square in the narrow ideal class group or not, the number of $\mathfrak{g}$ satisfying the assumption of Theorem 4.2, and the number of prime ideals $\mathfrak{p}$ of degree one above 3, respectively.

H. Boylan, *Jacobi Forms, Finite Quadratic Modules and Weil Representations over Number Fields*, Lecture Notes in Mathematics 2130,
DOI 10.1007/978-3-319-12916-7

Table A.1 Number fields K with $[K:\mathbb{Q}]=2$

D_K	s	f	∂_K	$\mathfrak{g}$	$\mathfrak{p}$
17	1	x^2-x-4	✓	1	0
41	1	x^2-x-10	✓	1	0
57	2	x^2-x-14		1	1
65	2	x^2-x-16	✓	1	0
73	4	x^2-x-18	✓	4	2
89	1	x^2-x-22	✓	1	0
97	4	x^2-x-24	✓	4	2
113	1	x^2-x-28	✓	1	0
129	2	x^2-x-32		1	1
137	1	x^2-x-34	✓	1	0
145	4	x^2-x-36	✓	2	2
185	2	x^2-x-46	✓	1	0
193	4	x^2-x-48	✓	4	2
201	2	x^2-x-50		1	1
217	4	x^2-x-54		2	2
233	1	x^2-x-58	✓	1	0
241	4	x^2-x-60	✓	4	2
257	1	x^2-x-64	✓	1	0
265	4	x^2-x-66	✓	2	2
273	4	x^2-x-68		1	1
281	1	x^2-x-70	✓	1	0
305	2	x^2-x-76	✓	1	0
313	4	x^2-x-78	✓	4	2
337	4	x^2-x-84	✓	4	2
353	1	x^2-x-88	✓	1	0
377	2	x^2-x-94	✓	1	0
401	1	$x^2-x-100$	✓	1	0
409	4	$x^2-x-102$	✓	4	2
417	2	$x^2-x-104$		1	1
433	4	$x^2-x-108$	✓	4	2

Table A.2 Number fields K with $[K : \mathbb{Q}] = 3$

D_K	s	f	∂_K	$\mathfrak{g}$	$\mathfrak{p}$
961	1	$x^3 - x^2 - 10x + 8$	✓	1	0
1,849	1	$x^3 - x^2 - 14x - 8$	✓	1	0
3,969	2	$x^3 - 21x - 28$	✓	2	1
4,481	2	$x^3 - 17x - 8$		1	1
7,057	1	$x^3 - x^2 - 22x + 32$	✓	1	0
7,441	1	$x^3 - x^2 - 22x - 16$	✓	1	0
8,281	1	$x^3 - x^2 - 30x + 64$	✓	1	0
8,289	2	$x^3 - 21x - 12$	✓	2	1
8,713	1	$x^3 - 25x - 32$	✓	1	0
9,153	2	$x^3 - 21x - 4$	✓	2	1
10,641	4	$x^3 - x^2 - 22x + 16$	✓	4	2
11,137	1	$x^3 - x^2 - 22x + 8$	✓	1	0
11,665	1	$x^3 - x^2 - 26x + 40$	✓	1	0
11,881	1	$x^3 - x^2 - 36x + 4$	✓	1	0
13,689	2	$x^3 - 39x - 26$	✓	2	1
14,129	2	$x^3 - x^2 - 26x - 16$		1	1
14,609	2	$x^3 - x^2 - 26x + 32$		1	1
15,641	2	$x^3 - 29x - 36$	✓	2	1
15,961	1	$x^3 - x^2 - 30x - 32$	✓	1	0
16,129	1	$x^3 - x^2 - 42x - 80$	✓	1	0
16,369	1	$x^3 - x^2 - 26x - 8$	✓	1	0
16,649	2	$x^3 - x^2 - 34x - 48$	✓	2	1
16,689	4	$x^3 - x^2 - 26x + 24$	✓	4	2
17,689	1	$x^3 - x^2 - 44x + 64$	✓	1	0
18,201	4	$x^3 - x^2 - 30x + 48$	✓	4	2
19,441	1	$x^3 - 37x - 68$	✓	1	0
20,073	4	$x^3 - x^2 - 30x - 24$	✓	4	2
20,385	2	$x^3 - 33x - 48$	✓	2	1
21,281	2	$x^3 - x^2 - 42x + 104$	✓	2	1
23,321	2	$x^3 - x^2 - 30x - 16$		1	1

Table A.3 Number fields K with $[K : \mathbb{Q}] = 4$

D_K	s	f	∂_K	$\mathfrak{g}$	$\mathfrak{p}$
122,825	2	$x^4 - x^3 - 23x^2 + x + 86$	✓	1	0
164,441	1	$x^4 - 2x^3 - 13x^2 + 14x + 32$	✓	1	0
171,377	1	$x^4 - 2x^3 - 19x^2 + 20x + 32$	✓	1	0
274,625	2	$x^4 - x^3 - 24x^2 + 4x + 16$	✓	1	0
282,353	1	$x^4 - x^3 - 35x^2 + 41x + 202$	✓	1	0
310,985	2	$x^4 - 2x^3 - 13x^2 + 14x + 8$	✓	1	0
314,721	2	$x^4 - 25x^2 + 16$	✓	1	0
317,033	1	$x^4 - 2x^3 - 17x^2 + 18x + 64$	✓	1	0
340,857	2	$x^4 - 2x^3 - 13x^2 + 14x + 16$	✓	1	0
356,337	2	$x^4 - x^3 - 37x^2 + 25x + 268$	✓	1	0
379,457	2	$x^4 - 2x^3 - 23x^2 + 24x + 76$	✓	1	0
389,017	1	$x^4 - x^3 - 27x^2 + 41x + 2$	✓	1	0
393,129	2	$x^4 - x^3 - 37x^2 + 97x + 4$	✓	1	0
393,329	1	$x^4 - x^3 - 39x^2 + 9x + 302$	✓	1	0
471,537	2	$x^4 - x^3 - 25x^2 + 25x + 64$	✓	1	0
485,809	1	$x^4 - 29x^2 + 36$	✓	1	0
500,033	1	$x^4 - 2x^3 - 21x^2 - 18x + 8$	✓	1	0
506,617	1	$x^4 - 2x^3 - 21x^2 + 22x + 104$	✓	1	0
532,521	4	$x^4 - x^3 - 27x^2 - 7x + 82$		1	1
624,529	1	$x^4 - 2x^3 - 27x^2 + 28x + 128$	✓	1	0
626,441	1	$x^4 - 21x^2 - 8x + 20$	✓	1	0
663,833	1	$x^4 - x^3 - 51x^2 + 49x + 514$	✓	1	0
668,457	2	$x^4 - 2x^3 - 33x^2 + 34x + 136$	✓	1	0
674,057	1	$x^4 - 23x^2 - 2x + 88$	✓	1	0
704,969	1	$x^4 - x^3 - 33x^2 - 39x + 8$	✓	1	0
751,409	1	$x^4 - 23x^2 - 6x + 80$	✓	1	0
754,769	1	$x^4 - x^3 - 26x^2 + 8x + 64$	✓	1	0
756,313	1	$x^4 - x^3 - 53x^2 + 33x + 596$	✓	1	0
768,713	1	$x^4 - 21x^2 - 4x + 32$	✓	1	0
830,297	4	$x^4 - x^3 - 57x^2 + x + 664$	✓	1	0
860,353	2	$x^4 - 2x^3 - 55x^2 + 56x + 172$	✓	1	0
906,593	1	$x^4 - 2x^3 - 31x^2 + 32x + 188$	✓	1	0
996,761	1	$x^4 - 2x^3 - 29x^2 + 30x + 208$	✓	1	0

Glossary

We list in roughly alphabetical order the basic notations which are used throughout this monograph.

$(\mathfrak{a}, \mathfrak{b})$	The greatest common divisor of the integral $\mathscr{O}_K$-ideals $\mathfrak{a}$ and $\mathfrak{b}$.
$\mathbb{C}^\infty(V)$	The space of functions which are differentiable for all degrees of differentiation defined on a $\mathbb{C}$-vector space V.
$\mathfrak{d}_K, \mathfrak{d}$	The different of the number field K.
dR	The principal ideal generated by the element d of the ring R.
$\dim_K V$	The dimension of the K-vector space V over the field K. If $K = \mathbb{C}$, we shortly write $\dim V$.
$e\{c\}$	The value $\exp\big(2\pi i \operatorname{tr}(c)\big)$, where c is an element of $\mathbb{C} \otimes_{\mathbb{Q}} K$ [1] (K a number field).
$\Gamma, \tilde{\Gamma}$	The group $\mathrm{SL}(2, \mathscr{O})$ and its metaplectic cover (see Sect. 3.3), respectively.
$\mathrm{GL}(V)$	The group of all automorphisms of a $\mathbb{C}$-vector space V.
$\mathbb{H}$	The upper half plane.
I	The element $(1, -1)$ in the metaplectic cover of $\mathrm{SL}(2, \mathscr{O})$.
$\mathrm{Hol}(V)$	The space of holomorphic functions of a $\mathbb{C}$-vector space V.
$\mathrm{Im}(\phi)$	The image of the map ϕ.
$\mathrm{Ker}(\phi)$	The kernel of the map ϕ.
μ	The Möbius μ-function, i.e. the multiplicative function μ on the semi-group of integral $\mathscr{O}_K$-ideals which for prime ideal power $\mathfrak{p}^n$ assumes the values $1, -1$ and 0 accordingly as $n = 0$ or $n = 1$ or $n \geq 2$.
$\precnsim_l$	The group of lth roots of unity.
$\precnsim_\infty$	The group of all roots of unity.
$\mathrm{N}_{K/\mathbb{Q}}(a)$	The norm of an element a in the number field K.

[1]For the definition of the trace of an element in $\mathbb{C} \otimes_{\mathbb{Q}} K$, we refer to Sect. 3.2.

H. Boylan, *Jacobi Forms, Finite Quadratic Modules and Weil Representations over Number Fields*, Lecture Notes in Mathematics 2130,
DOI 10.1007/978-3-319-12916-7

$\mathrm{N}(\mathfrak{a})$	The norm of the ideal $\mathfrak{a}$.
$\mathcal{O}_K, \mathcal{O}$	The ring of integers of the number field K.
q^t	The function on $\mathcal{H}^2$ defined by $e\{t\tau\}$.
R^*	The invertible elements of the ring R under multiplication.
R^n	The R-module of column vectors of length n over the ring R.
S	The matrix $\left(\begin{smallmatrix} 0 & -1 \\ 1 & 0 \end{smallmatrix}\right)$.
$\mathrm{SL}(n, R)$	The subgroup of elements of $\mathrm{GL}(n, R)$ which have determinant 1, where R is a ring.
$\mathbb{S}^1$	The group of all complex numbers whose absolute value equals one.
$\mathrm{Stab}(x)$	The stabilizer of x under a given group action.
$\sigma_0(\mathfrak{a})$	The number of ideals dividing the integral $\mathcal{O}_K$-ideal $\mathfrak{a}$.
T_b	The matrix $\left(\begin{smallmatrix} 1 & b \\ 0 & 1 \end{smallmatrix}\right)$, where b is an element of the ring R.
$\mathrm{tr}_{K/\mathbb{Q}}(a)$	The trace of an element a in the number field K.
$\sqrt{z}$	The root of $z \in \mathbb{C}^*$ whose argument lies in the interval $(\frac{-\pi}{2} \frac{\pi}{2}]$.

[2]Here $\mathcal{H}$ is a the upper half plane in $\mathbb{C} \otimes_{\mathbb{Q}} K$ (K a number field) as defined in Sect. 3.2.

References

[BHS14] H. Boylan, S. Hayashida, N.-P. Skoruppa, On the computation of Jacobi forms over totally real number fields with an explicit example over $\mathbb{Q}(\sqrt{5})$ (2014, preprint)

[Bor95] Bordeaux Computational Number Theory Group, The number field tables (1995). http://pari.math.u-bordeaux.fr/pub/pari/packages/nftables.tgz

[Boy14] H. Boylan, Jacobi Eisenstein series over number fields. (2014, preprint)

[BS13] H. Boylan, N.-P. Skoruppa, Linear characters of SL_2 over Dedekind domains. J. Algebra **373**, 120–129 (2013)

[BS14a] H. Boylan, N.-P. Skoruppa, Explicit formulas for Weil representations of SL(2). (2014, preprint)

[BS14b] H. Boylan, N.-P. Skoruppa, Jacobi forms over number fields. (2014, in preparation)

[Ebe02] W. Ebeling, *Lattices and Codes*. Advanced Lectures in Mathematics, revised edn. (Friedr. Vieweg & Sohn, Braunschweig, 2002). [A course partially based on lectures by F. Hirzebruch]

[EZ85] M. Eichler, D. Zagier, *The Theory of Jacobi Forms*. Progress in Mathematics, vol 55 (Birkhäuser Boston Inc., Boston, 1985)

[FH91] W. Fulton, J. Harris, *Representation Theory*. Graduate Texts in Mathematics, vol 129 (Springer, New York, 1991) [A first course, Readings in Mathematics]

[Fre90] E. Freitag, *Hilbert Modular Forms* (Springer, Berlin, 1990)

[FT93] A. Fröhlich, M.J. Taylor, *Algebraic Number Theory*. Cambridge Studies in Advanced Mathematics, vol 27 (Cambridge University Press, Cambridge, 1993)

[GKZ87] B. Gross, W. Kohnen, D. Zagier, Heegner points and derivatives of L-series. II. Math. Ann. **278**(1–4), 497–562 (1987)

[Hec81] E. Hecke, *Lectures on the Theory of Algebraic Numbers*. Graduate Texts in Mathematics, vol 77 (Springer, New York, 1981) [Translated from the German by George U. Brauer, Jay R. Goldman and R. Kotzen]

[Kub67] T. Kubota, Topological covering of SL(2) over a local field. J. Math. Soc. Jpn. **19**, 114–121 (1967)

[MH73] J. Milnor, D. Husemoller, *Symmetric Bilinear Forms* (Springer, New York, 1973) [Ergebnisse der Mathematik und ihrer Grenzgebiete, Band 73]

[Neu99] J. Neukirch, *Algebraic Number Theory*. Grundlehren der Mathematischen Wissenschaften [Fundamental Principles of Mathematical Sciences], vol 322 (Springer, Berlin, 1999) [Translated from the 1992 German original and with a note by Norbert Schappacher, With a foreword by G. Harder]

H. Boylan, *Jacobi Forms, Finite Quadratic Modules and Weil Representations over Number Fields*, Lecture Notes in Mathematics 2130,
DOI 10.1007/978-3-319-12916-7

[Ser73] J.-P. Serre, *A Course in Arithmetic* (Springer, New York, 1973) [Translated from the French, Graduate Texts in Mathematics, No. 7]

[Ser77] J.-P. Serre, *Linear Representations of Finite Groups* (Springer, New York, 1977) [Translated from the second French edition by Leonard L. Scott, Graduate Texts in Mathematics, vol. 42]

[Sko85] N.-P. Skoruppa, Über den Zusammenhang zwischen Jacobiformen und Modulformen halbganzen Gewichts. Bonner Mathematische Schriften [Bonn Mathematical Publications], 159. Universität Bonn Mathematisches Institut, Bonn, 1985. Dissertation, Rheinische Friedrich-Wilhelms-Universität, Bonn (1984)

[Sko10] N.-P. Skoruppa, Finite quadratic modules, weil representations and vector valued modular forms. Notes of courses given at Universität Siegen, Harish-Chandra Research Institute and Durham University (2010)

[SS14] N.-P. Skoruppa, F. Strömberg, Dimension formulas for vector-valued Hilbert modular forms (2014, in preparation)

[SW14] N.-P. Skoruppa, L. Walling, Hecke operators for Jacobi forms over number fields (2014, in preparation)

[O'M00] O. Timothy O'Meara, *Introduction to Quadratic Forms*. Classics in Mathematics (Springer, Berlin, 2000) [Reprint of the 1973 edition]

[Vas72] L.N. Vaseršteĭn, The group SL_2 over Dedekind rings of arithmetic type. Mat. Sb. (N.S.) **89**(131), 313–322, 351 (1972)

[vdG88] G. van der Geer, *Hilbert Modular Surfaces*. Ergebnisse der Mathematik und ihrer Grenzgebiete (3) [Results in Mathematics and Related Areas (3)] (Springer, Berlin, 1988)

LECTURE NOTES IN MATHEMATICS Springer

Edited by J.-M. Morel, B. Teissier; P.K. Maini

Editorial Policy (for the publication of monographs)

1. Lecture Notes aim to report new developments in all areas of mathematics and their applications - quickly, informally and at a high level. Mathematical texts analysing new developments in modelling and numerical simulation are welcome.
   Monograph manuscripts should be reasonably self-contained and rounded off. Thus they may, and often will, present not only results of the author but also related work by other people. They may be based on specialised lecture courses. Furthermore, the manuscripts should provide sufficient motivation, examples and applications. This clearly distinguishes Lecture Notes from journal articles or technical reports which normally are very concise. Articles intended for a journal but too long to be accepted by most journals, usually do not have this "lecture notes" character. For similar reasons it is unusual for doctoral theses to be accepted for the Lecture Notes series, though habilitation theses may be appropriate.

2. Manuscripts should be submitted either online at www.editorialmanager.com/lnm to Springer's mathematics editorial in Heidelberg, or to one of the series editors. In general, manuscripts will be sent out to 2 external referees for evaluation. If a decision cannot yet be reached on the basis of the first 2 reports, further referees may be contacted: The author will be informed of this. A final decision to publish can be made only on the basis of the complete manuscript, however a refereeing process leading to a preliminary decision can be based on a pre-final or incomplete manuscript. The strict minimum amount of material that will be considered should include a detailed outline describing the planned contents of each chapter, a bibliography and several sample chapters.
   Authors should be aware that incomplete or insufficiently close to final manuscripts almost always result in longer refereeing times and nevertheless unclear referees' recommendations, making further refereeing of a final draft necessary.
   Authors should also be aware that parallel submission of their manuscript to another publisher while under consideration for LNM will in general lead to immediate rejection.

3. Manuscripts should in general be submitted in English. Final manuscripts should contain at least 100 pages of mathematical text and should always include

   – a table of contents;
   – an informative introduction, with adequate motivation and perhaps some historical remarks: it should be accessible to a reader not intimately familiar with the topic treated;
   – a subject index: as a rule this is genuinely helpful for the reader.

   For evaluation purposes, manuscripts may be submitted in print or electronic form (print form is still preferred by most referees), in the latter case preferably as pdf- or zipped ps-files. Lecture Notes volumes are, as a rule, printed digitally from the authors' files. To ensure best results, authors are asked to use the LaTeX2e style files available from Springer's web-server at:

   ftp://ftp.springer.de/pub/tex/latex/svmonot1/ (for monographs) and
   ftp://ftp.springer.de/pub/tex/latex/svmultt1/ (for summer schools/tutorials).

Additional technical instructions, if necessary, are available on request from lnm@springer.com.

4. Careful preparation of the manuscripts will help keep production time short besides ensuring satisfactory appearance of the finished book in print and online. After acceptance of the manuscript authors will be asked to prepare the final LaTeX source files and also the corresponding dvi-, pdf- or zipped ps-file. The LaTeX source files are essential for producing the full-text online version of the book (see http://www.springerlink.com/openurl.asp?genre=journal&issn=0075-8434 for the existing online volumes of LNM). The actual production of a Lecture Notes volume takes approximately 12 weeks.

5. Authors receive a total of 50 free copies of their volume, but no royalties. They are entitled to a discount of 33.3 % on the price of Springer books purchased for their personal use, if ordering directly from Springer.

6. Commitment to publish is made by letter of intent rather than by signing a formal contract. Springer-Verlag secures the copyright for each volume. Authors are free to reuse material contained in their LNM volumes in later publications: a brief written (or e-mail) request for formal permission is sufficient.

Addresses:

Professor J.-M. Morel, CMLA,
École Normale Supérieure de Cachan,
61 Avenue du Président Wilson, 94235 Cachan Cedex, France
E-mail: morel@cmla.ens-cachan.fr

Professor B. Teissier, Institut Mathématique de Jussieu,
UMR 7586 du CNRS, Équipe "Géométrie et Dynamique",
175 rue du Chevaleret
75013 Paris, France
E-mail: teissier@math.jussieu.fr

For the "Mathematical Biosciences Subseries" of LNM:

Professor P. K. Maini, Center for Mathematical Biology,
Mathematical Institute, 24-29 St Giles,
Oxford OX1 3LP, UK
E-mail: maini@maths.ox.ac.uk

Springer, Mathematics Editorial, Tiergartenstr. 17,
69121 Heidelberg, Germany,
Tel.: +49 (6221) 4876-8259

Fax: +49 (6221) 4876-8259
E-mail: lnm@springer.com